Green Finance and Investment

Financing Climate Action in Eastern Europe, the Caucasus and Central Asia

Please cite this publication as:
OECD (2016), *Financing Climate Action in Eastern Europe, the Caucasus and Central Asia*, Green Finance and Investment, OECD Publishing, Paris.
http://dx.doi.org/10.1787/9789264266339-en

ISBN 978-92-64-26632-2 (print)
ISBN 978-92-64-26633-9 (PDF)

Series: Green Finance and Investment
ISSN 2409-0336 (print)
ISSN 2409-0344 (online)

Photo credits: Cover © mika48/Shutterstock.com.

Corrigenda to OECD publications may be found on line at: *www.oecd.org/about/publishing/corrigenda.htm*.

Foreword

The 21st session of the Conference of the Parties to the United Nations Framework Convention on Climate Change (COP21) adopted an historic climate agreement in Paris. Achieving the goals under the Paris Agreement hinges on implementation of the countries' nationally determined contributions and various mechanisms under the Convention through greater investment. A drastic shift in finance that flows from "brown" to "green" will be essential at both global and national levels.

This report aims to shed light on how the countries of Eastern Europe, the Caucasus and Central Asia (EECCA) have been working with development co-operation partners to finance climate actions. It explores how the countries can assess and improve their readiness to seize further opportunities to access various climate finance sources in the future. The EECCA countries studied are Armenia, Azerbaijan, Belarus, Georgia, Kazakhstan, Kyrgyzstan, Moldova, Tajikistan, Turkmenistan, Ukraine and Uzbekistan. Most of the EECCA countries communicated their intended nationally determined contributions (INDCs) in time for COP21, along with their mitigation and/or adaptation targets. Finance is an indispensable means of implementation to achieve the climate targets of the EECCA countries.

This report was prepared as part of the project "International Climate Finance for EECCA", supported by the German Federal Ministry for the Environment, Nature Conservation, Building and Nuclear Safety and implemented under the GREEN Action Programme hosted by the Organisation for Economic Co-operation and Development (OECD). The project focuses on climate finance readiness of the EECCA countries to access and use climate finance from various international sources effectively. The report benefitted from the discussions at the Expert Workshop on International Climate Finance for EECCA that was held on 11 July 2016 in Paris.

Chapter 1 provides a background, aims and key policy recommendations of the report. Chapter 2 presents a landscape of climate-related development finance at the regional level in 2013 and 2014. Chapter 3 outlines key issues on access of the EECCA countries to scaled-up finance for their climate actions. Finally chapter 4 summaries the country-level analyses of climate-related development finance in the 11 EECCA countries. The report is accompanied by 11 reports for all the EECCA countries that contain more detailed analysis of climate-related development finance flows, their INDCs, and the policies and institutional arrangements of the countries studied. Country reports can be accessed at www.oecd.org/env/outreach/eap-tf.htm. Multiple sources informed this report and the 11 country-level reports, including the OECD Creditor Reporting System (CRS) and project-related documents from a range of providers of development finance.

Takayoshi Kato (OECD Environment Directorate) led drafting and analysis with substantive inputs and thorough review by Nelly Petkova (OECD Environment Directorate) under the guidance of Kumi Kitamori (Head of the Green Growth and Global Relations Division, Environment Directorate, OECD). The authors gratefully acknowledge insights provided at the July 2016 workshop by participants from Armenia, Belarus, Georgia, Kazakhstan, Moldova, Tajikistan, Ukraine, Uzbekistan, the European Commission, the European Bank for Reconstruction and Development, the European Investment Bank, Gesellschaft für Internationale Zusammenarbeit (GIZ), the United Nations Environment Programme and the REN21 Secretariat. The report also greatly benefited from expert review and valuable input from colleagues at the OECD Secretariat: Kumi Kitamori, Juan Casado Asensio, Naeeda Crishna Morgado, Raphaël Jachnik, Alexandre Martoussevitch, Krzysztof Michalak, Mariana Mirabile and Mikaela Rambali, as well as external experts: Marko Berglund (UN Environment), Maya Eralieva (GIZ Kyrgyzstan), Maria Falaleeva (EKAPRAEKT), Martin Hullin (REN21), Natalia Kushko (USAID Municipal Energy Reform Project), Nino Lazahvili (Georgia), Dorit Lehr (GIZ), Veronica Lopotenco (Moldova), Zafar Makhmadov (Tajikistan), Marta Modelewska (EBRD) and Laura Würtenberger (GIZ). The authors would also like to thank Lupita Johanson, Douglas Herrick and Janine Treves (OECD) who assisted with the processing of the publication and Maria Dubois and Shukhrat Ziyaviddinov (OECD) who provided administrative support.

Table of contents

Figures

Tables

Boxes

Reader's guide

Data sources

Analysis in this report draws on multiple information sources. Data on development finance flows were obtained from the OECD Creditor Reporting System (CRS) Aid Activities Database,[1] which is maintained by the OECD Development Assistance Committee (DAC). Information on climate targets and policies was retrieved from publicly available national policy documents, which directly and indirectly relate to climate actions (i.e. climate change mitigation and adaptation). The data sources also include a number of project-level documents prepared by bilateral and multilateral providers of finance. These include members of the OECD DAC, multilateral development banks (MDBs) and international climate funds, as well as some South-South co-operation and non-DAC member contributions.

This study tracks development finance flows for activities that target climate mitigation or adaptation as either their principle objective or significant objective. Data on development finance flows in this report are based on the OECD DAC CRS, and cover 2013 and 2014. This database allows for an approximate quantification of climate-related development finance flows that target climate mitigation and adaptation objectives of the Rio Conventions.

The OECD DAC CRS records face values of activities on the dates when grant or loan agreements are signed with recipients (i.e. commitment, but not disbursement). Therefore, there may be gaps between results from the DAC CRS and recipient countries' external climate-related development finance statistics on the ground, especially when observed over a longer period. Data sources are limited to the OECD DAC member countries, the MDBs and climate funds that report to the DAC CRS. Therefore, sources do not include some non-DAC member donors such as the People's Republic of China and the Russian Federation, or private sector financing, which are likely to have provided a significant amount of finance to some EECCA countries. The OECD DAC tracks, monitors and/or estimates development finance of countries beyond the DAC members, but activity-level data on climate-related development finance from these countries is not available as of August 2016. The OECD also hosts the Research Collaborative on Tracking Private Climate Finance, aiming to develop methodologies for identifying private finance for climate action in developing counties, and more specifically for estimating publicly-mobilised private climate finance.

Cut-off date

The landscape of climate finance and climate-related policies are rapidly evolving in many countries. The cut-off date for inclusion of policy developments in this report was August 2016.

Note

1. For more details, see www.oecd.org/dac/stats/climate-change.htm and on the DAC members see www.oecd.org/dac/dacmembers.htm.

Abbreviations and acronyms

ADB	Asian Development Bank
AF	The Adaptation Fund
BAU	Business as Usual
bln	Billion
BMUB	The Federal Ministry for the Environment, Nature Conservation, Building and Nuclear Safety (of Germany)
BMZ	The Federal Ministry for Economic Cooperation and Development (of Germany)
CEP	Committee on Environmental Protection (of Tajikistan)
CIF	Climate Investment Funds
CPI	Climate Policy Initiative
CRS	The OECD Creditor Reporting System
DAC	The OECD Development Assistance Committee
EBRD	European Bank for Reconstruction and Development
EC	European Commission
EECCA	Eastern Europe, the Caucasus and Central Asia
EIB	European Investment Bank
ENP	The European Neighbourhood Policy
EU	European Union
EUR	Euros
GCF	Green Climate Fund
GDP	Gross domestic product
GEDF	Georgian Energy Development Fund
GEF	Global Environment Facility
GHG	Greenhouse gas
GIZ	Deutsche Gesellschaft für Internationale Zusammenarbeit GmbH
GNI	Gross national income
IBRD	International Bank for Reconstruction and Development
IDA	International Development Association

IEA	International Energy Agency
IFC	International Finance Corporation
IKI	The International Climate Initiative (of BMUB)
INDC	Intended Nationally Determined Contribution
ITF	International Transport Forum at the OECD
LEDS	Low-Emission Development Strategies
LULUCF	Land use, land-use change and forestry
MDB	Multilateral development bank
mln	millions
MWh	megawatt hours
NAMA	Nationally Appropriate Mitigation Action
NAP	National Adaptation Plan
NDA	National Designated Authority
NDC	Nationally Determined Contribution
NEA	Nuclear Energy Agency (within the OECD)
NIE	National Implementing Entity
NIF	Neighbourhood Investment Facility of the EU
ODA	Official Development Assistance
OECD	Organisation for Economic Co-operation and Development
PPCR	The Pilot Program for Climate Resilience
PPP	Purchasing power parity
R2E2	The Renewable Resources and Energy Efficiency Fund (of Armenia)
RIE	Regional Implementing Entity
SDGs	Sustainable Development Goals
SOFAZ	State Oil Fund of Azerbaijan
TNA	Technology Needs Assessment
TWN	Third World Network
UNDP	United Nations Development Programme
UNECE	United Nations Economic Commission for Europe
UNEP	United Nations Environment Programme
UNFCCC	United Nations Framework Convention on Climate Change
USD	US dollars
VAT	Value added tax
WB	World Bank
WRI	World Resources Institute

Executive summary

Finance is crucial to enable an effective and progressive global response to the urgent threat of climate change. Bilateral and multilateral development co-operation partners committed USD 3.3 billion of development finance per year over 2013 and 2014 to climate actions (mitigation or adaptation, or both) in 11 countries of Eastern Europe, the Caucasus and Central Asia (EECCA) – Armenia, Azerbaijan, Belarus, Georgia, Kazakhstan, Kyrgyzstan, Moldova, Tajikistan, Turkmenistan, Ukraine and Uzbekistan. Compared to other regions in the world and in terms of gross domestic product (GDP) per capita levels, the commitment to these countries represented a relatively fair share of development finance for their climate actions.

This report focuses on development finance from international public sources, which targets climate actions in developing countries (hereafter "climate-related development finance"). The report aims to:

- improve clarity on how the 11 EECCA countries and their development co-operation partners have been working together to finance climate actions (mitigation and adaptation) in the countries
- explore how the EECCA countries can assess and identify areas for improvement in their readiness to seize further opportunities to access various climate finance sources in the future.

While significant amounts of climate-related development finance were committed to the EECCA countries in 2013 and 2014, the scale of finance directed to each country is considerably different. Nevertheless, all the EECCA countries still need scaled-up finance from international and domestic sources for enhanced climate actions to achieve their mitigation and adaptation targets. The Paris Agreement stresses the importance of thematic balance between mitigation and adaptation finance. Yet mitigation finance (81%) outweighs adaptation finance (11%) in the EECCA region; 8% of finance targets both mitigation and adaptation. While such thematic imbalance is observed globally, its difference is greater in the EECCA than the global average.

In the EECCA region, nearly half of climate-related development finance in 2013 and 2014 was committed to the energy sector. This high share reflects investment needs to rehabilitate and replace ageing energy-related infrastructure on both supply and demand sides. Despite the number of energy- and climate-related strategies and policies that have helped improve energy efficiency, the energy intensity of the EECCA economies is still significantly higher than that of OECD member countries. In most of the EECCA countries, the share of renewable energy in the energy mix remains low. Some countries have developed secondary legislation and regulations of national-level energy. However, the impact of these actions has not yet triggered domestic and international investment in low-carbon development of the energy sector.

Adaptation policies have been developed less well than mitigation policies in most of the EECCA countries, resulting in challenges for accessing adaptation finance. A much smaller amount of finance is directed to adaptation even in countries such as Belarus, Kyrgyzstan and Georgia, whose INDCs explicitly mention that adaptation is priority. There is great potential to mainstream climate considerations, particularly on adaptation, into development finance and policy planning in the EECCA countries. For instance, climate-related development finance committed to the agriculture and forestry sector accounts for one-third of total development finance for this sector, much lower than the global average.

Both bilateral and multilateral development co-operation partners play important roles in delivering climate-related development finance to the EECCA countries; multilateral channels delivered nearly two-thirds of total climate-related development finance in 2013 and 2014. Multilateral development banks (MDBs) mainly provide non-concessional loans, which are the most used instruments to deliver climate-related development finance to the EECCA countries. Bilateral channels and dedicated climate funds are important providers of grants and concessional loans through, for instance, technical assistance, project investments and co-financing to projects supported by MDBs.

"Pathways" in terms of both actions and finance towards the EECCA countries' climate targets are still at an early stage of development. Armenia, Georgia, Kazakhstan, Kyrgyzstan, Moldova, Tajikistan and Turkmenistan have established stricter mitigation targets that are conditional on international support (e.g. finance and technologies). But only Kyrgyzstan and Moldova have clarified how much financing will be needed to achieve such targets in the INDCs. Apart from the INDCs, the EECCA countries have developed many climate policies and other types of policies indirectly related to climate actions. Yet many of these policies would further benefit from more concrete planning for implementation and better coherence among them. Armenia, Azerbaijan, Georgia, Kazakhstan, Moldova, Turkmenistan and Uzbekistan have identified technology needs and associated costs. However, concrete investment planning based on results from these Technology Needs Assessments (TNAs) has not yet been well developed.

Country ownership over implementing climate and other development actions has been an important element for pursuing sustainable development. Direct access modalities allow a country's national institution to directly access financial resources of specific climate funds such as the Green Climate Fund (GCF) and the Adaptation Fund, and are gaining increasing traction as a means to enhancing country ownership and efficiency. None of the EECCA countries has accessed such climate funds using direct access modalities to date. While recognising the important role of international entities in delivering climate finance, some EECCA countries may benefit from pursuing direct access to improve efficiency (e.g. reducing transaction costs) and strengthen their ownership over accessing and absorbing international finance. Indeed some of the countries, such as Uzbekistan, have started to prepare for direct access.

The EECCA countries and their development co operation partners have been engaging in a range of work, spanning from analysis and technical assistance to infrastructure investments. Work under the OECD-hosted GREEN Action Programme intends to complement and contribute to activities on climate change in the EECCA region. In future, the EECCA countries and their development co-operation partners could pursue work such as the following:

- taking stock of climate and non-climate targets and policies to explore whether they are coherent and mutually reinforcing, and how such policy alignment can be further improved

- examining actionable steps to ensure the effective and continuous implementation of measures towards their INDCs, including investment planning for the measures
- assessing and improving key institutions' capacity to develop a pipeline of climate-related projects and access necessary finance, including emerging sources such as the GCF
- improving transparency about climate finance flows to provide a robust basis for domestic discussions and strengthen trust with providers of support, aiming to further mobilise finance for climate action from domestic and international sources
- exploring better use of national public institutions (e.g. national funding entities) for mobilising climate finance and improving country ownership over accessing and using financial resources.

Chapter 1

Scaling-up finance to support climate actions in EECCA countries

This chapter provides an overview of 11 countries of Eastern Europe, the Caucasus and Central Asia (EECCA), which are the scope of this report, in terms of their economies, populations and climate change related targets. This chapter also summarises the intended nationally determined contributions that the EECCA countries submitted in time for the 21st Conference of the Parties (COP21) to the United Nations Framework Convention on Climate Change (UNFCCC). Finally, this chapter outlines a set of recommendations for policy development that the EECCA countries and their development co-operation partners could pursue to scale-up finance for climate finance in the region, based on the results of the analysis detailed in Chapters 2, 3 and 4.

Background

Scientific evidence warrants strong action to mitigate greenhouse gas (GHG) emissions around the world to reduce the risk of irreversible impacts of climate change on ecosystems, societies and economies. At the same time, vulnerable populations are already experiencing the impacts of a changing climate. Thus, actions that strengthen adaptation to such adverse effects of climate change are also needed. In 2015, the 21st Conference of the Parties (COP21) to the United Nations Framework Convention on Climate Change (UNFCCC) adopted the Paris Agreement, while the United Nations General Assembly adopted the 2030 Agenda for Sustainable Development and its 17 Sustainable Development Goals (SDGs). To strengthen the global response to the threat of climate change in the context of sustainable development, the Paris Agreement set three goals towards low-carbon and climate-resilient development as shown below:

- Hold the global average surface temperature increase to well below 2°C and pursue efforts to limit it to 1.5°C above pre-industrial levels.
- Increase the ability to adapt to adverse climate impacts and foster resilience and low greenhouse gas (GHG) development without threatening food production.
- Make finance flows consistent with a pathway towards low GHG emissions and climate-resilient development.

Achieving these goals hinges on implementation of the countries' nationally determined contributions and various mechanisms under the Convention. Such implementation requires investment. There is no shortage of available capital globally (OECD, 2015b), but a drastic shift in finance that flows from "brown" to "green" will be essential at the global and national levels.

A large amount of finance has already been mobilised from domestic, international, private and public sources to pursue low-carbon and climate-resilient development around the world. However, finance needs to be substantially scaled up from the current level. For instance, the Standing Committee on Finance under the UNFCCC estimates that the global total climate finance, which includes public and private financial resources devoted to addressing climate change, ranged from USD 340-650 billion in 2014 in all countries (UNFCCC, 2016). The finance needed to achieve the mitigation and adaptation goals of the Paris Agreement, however, is much larger. For example, the International Energy Agency (IEA) estimates the costs of the full implementation of climate pledges expressed in the intended nationally determined contributions (INDCs) by 150 countries. This IEA analysis shows that approximately USD 13.5 trillion will be needed to achieve these INDC targets in the energy sector alone for energy efficiency and other low-carbon technologies from 2015-30, while the targets are not collectively sufficient to achieve the 2°C goal of the Paris Agreement (IEA, 2015).

This report focuses on international climate-related development finance from public sources directed to the 11 countries of Eastern Europe, the Caucasus and Central Asia (EECCA), which cover a large geographical area with total population of approximately 140 million. These countries are Armenia, Azerbaijan, Belarus, Georgia, Kazakhstan, Kyrgyzstan, Moldova, Tajikistan, Turkmenistan, Ukraine and Uzbekistan. The EECCA countries cover a large geographical area with total population of approximately 140 million and are markedly diverse in terms of geographical and population sizes, levels of economic development, political and economic structures, energy mixes, geopolitical circumstances and vulnerabilities to climate change (Table 1.1).

A number of similarities, however, do exist, including issues relating to climate change. For instance, energy-related infrastructure (for generation, transmission and consumption) in several EECCA countries was built when the countries were part of the Soviet Union, and has not been maintained or modernised. This has made such infrastructure among the most energy-inefficient in the world. In Kazakhstan, one of the wealthiest countries in the region, about 50% of energy-related infrastructure is more than 30 years-old; some installations are 50-60 years old and coal-based. The World Bank estimates that Uzbekistan lost USD 2 billion in 2011 – or 4.5% of its gross domestic product (GDP) – due to inefficient electricity transmission. These examples illustrate the significant level of financing needed for upgrading the energy sector infrastructure on both supply and demand sides. The share of renewable energy in the total energy supply is low in many of the EECCA countries where hydropower is not the major source for energy supply.

Kyrgyzstan, Moldova, Tajikistan and Uzbekistan are particularly vulnerable and have relatively low adaptive capacities, while the other countries are also likely to be negatively affected by a changing climate. The most vulnerable sectors in the region include agriculture, energy, water, disaster risk management, healthcare, forestry and biodiversity protection, although the needs and priorities vary among different countries. Even in countries with rich water resources, climate change may have significant negative consequences to their economies. For instance, irrigated areas in Armenia dropped in half between the 1980s and 2014 – from 300 000 ha to 150 000 ha , which may be partly due to a changing climate (Melkonyan, 2015).

Table 1.1. **Basic information on EECCA economies and CO2 emission levels (2013)**

	2014 – GDP PPP (constant 2011 international $ mln)	2014 – Population	2014 – GDP per capita, PPP (constant 2011 international $)	CO_2 emissions from fuel combustion ($MtCO_{2e}$)	CO_2 emissions (from fuel combustion) per unit of GDP PPP (kg CO_{2e} per $ of GDP)
Armenia	23 143.5	3 006 154	7 698.7	5.24	0.23
Azerbaijan	159 379.9	9 537 823	16 710.3	29.45	0.18
Belarus	164 292.8	9 470 000	17 348.8	58.25	0.35
Georgia	32 579.9	4 504 100	7 233.4	6.63	0.20
Kazakhstan	399 619.4	17 289 111	23 113.9	244.89	0.61
Kyrgyzstan	18 491.3	5 834 200	3 169.5	8.88	0.48
Moldova	16 905.5	3 556 400	4 753.5	6.70	0.40
Tajikistan	21 295.7	8 295 840	2 567.0	3.31	0.16
Turkmenistan	78 345.5	5 307 188	14 762.1	66.02	0.84
Ukraine	375 018.1	45 362 900	8 267.1	265.05	0.71
Uzbekistan	163 534.8	30 757 700	5 316.9	96.16	0.59
OECD average	1 378 277.0	36 599 903	37 657.9	343.93	0.25

Note: PPP stands for purchasing power parity.

Source: IEA (2016); World Bank (2016).

Intended nationally determined contributions (INDCs) of the EECCA countries

All EECCA countries, except Uzbekistan, submitted their INDCs to the UNFCCC Secretariat prior to COP21, with quantitative targets for reducing GHG emissions by 2030 or 2050 (Table 1.2). Seven INDCs (Armenia, Belarus, Georgia, Kyrgyzstan, Moldova, Tajikistan and Turkmenistan) also include sections on adaptation plans and/or actions, although their contents, format and timeframes differ significantly. Seven countries (Armenia, Georgia,

Kazakhstan, Kyrgyzstan, Moldova, Tajikistan and Turkmenistan) have established stricter mitigation targets that are conditional on international support (e.g. finance, technology and capacity building). However, only two INDCs (Kyrgyzstan and Moldova) explicitly indicate financial needs for implementing their climate actions. Georgia's INDC has communicated the long-term cost only for adaptation. Armenia's INDC does not estimate financial needs, but mentions possible domestic and international sources to achieve its mitigation and adaptation goals, as well as capacity building needs.

Half of the countries (Armenia, Azerbaijan, Kyrgyzstan, Tajikistan and Turkmenistan) explicitly prioritised their climate actions in their INDCs. Others (e.g. Kazakhstan, Georgia, Ukraine and Belarus) refer to existing national policy documents (such as Kazakhstan's Concept for Transition to a Green Economy) or their National Communications submitted to the UNFCCC, which outline the individual countries' priorities. Several INDCs refer to policy documents under development, which aim to determine specific climate adaptation and mitigation actions and priorities. Such policy documents include Nationally Appropriate Mitigation Actions (e.g. Armenia, Georgia and Turkmenistan), National Adaptation Plans (e.g. Armenia, Moldova and Turkmenistan) and Low-Emission Development Strategies under the UNFCCC system (e.g. Georgia and Moldova), as well as the countries' national policy frameworks.

Examining whether and how the EECCA's INDCs can help the countries implement concrete actions to achieve their long-term climate goals is beyond the scope of this report, but could be an important next step. Across the world, INDCs in many other developing countries tend to be linked with national priorities, including development goals and sectoral plans, and often highlight options for domestic finance sources and international support (Moore, 2015). In the long run, the EECCA countries may use the INDCs (or nationally determined contributions [NDCs][1]) as a basis for domestic discussions on overarching, long-term green investment planning. They can also be used to communicate – both domestically and internationally – national priorities and needs for finance and action. If this is the case, it would be useful to explore which elements of the INDCs (or NDCs) of the EECCA countries could be further improved and how.

Table 1.2. **Summary of INDCs by ODA-eligible EECCA countries**

	Mitigation component				
	Unconditional	Conditional on international support	Target year	Adaptation component included?	Quantified need for support mentioned?
Armenia	Limit emissions to an aggregate 633 million tCO_2e (5.4 tCO_2e per capita annually)	Achieve ecosystem-neutral GHG emissions in 2050 (2.07 tCO_2e per capita annually)	2050	Yes	No
Azerbaijan	Reduce GHG emissions by 35% from 1990 levels	-	2030	No	No
Belarus	Reduce GHG emissions level by 28% from 1990 levels	-	2030	Yes	No
Georgia	Reduce GHG emissions by 15% (excl. land use & forestry) below business as usual (BAU) levels	Reduce GHG emissions by 25% (excl. land use & forestry) below business as usual (BAU) levels	2030	Yes	No

Table 1.2. **Summary of INDCs by ODA-eligible EECCA countries** *(continued)*

	Mitigation component				
	Unconditional	Conditional on international support	Target year	Adaptation component included?	Quantified need for support mentioned?
Kazakhstan	Reduce GHG emissions by 15% from 1990 levels	Reduce GHG by 25% from 1990 level	2030	No	No
Kyrgyzstan	Reduce GHG emissions by between 11.49% and 13.75% below BAU levels	Reduce GHG emissions by between 29% and 30.89% below BAU levels	2030	Yes	Yes
Moldova	Reduce GHG emissions by 64-67% from 1990 levels	Reduce GHG emissions by 78 % from 1990 levels	2030	Yes	Yes
Tajikistan	Limit GHG or CO_2 emissions to 80-90% of 1990 levels	Limit GHG or CO_2 emissions to 65-75% of 1990 levels	2030	Yes	No
Turkmenistan	No growth in GHG emissions, or even reducing emissions (primarily by domestic sources but also with support of international finance)		2030	Yes	No
Ukraine	Reduce GHG emissions by 40% from 1990 levels	-	2030	No	No

Source: UNFCCC (2015).

Aims and structure of the report

This report focuses on climate-related development finance from bilateral and multilateral public-sector sources to the 11 EECCA countries with the aim of the following:

- improving clarity on how the 11 EECCA countries and their development co-operation partners worked together to finance climate actions during recent years
- exploring how the EECCA countries can assess and identify areas for improvement in their readiness to seize further opportunities to access various climate finance sources in the future.

The report does not provide a complete picture of climate finance flows from all possible sources (e.g. private-sector investments are not included) due largely to limited data availability.

The report intends to facilitate a better understanding of international development finance flows for climate actions in the EECCA region in terms of sectors/areas, providers and financing structures for individual projects, as well as domestic institutions involved in accessing and using such finance. It consists of the following parts:

- Chapter 2. Landscape of climate-related development finance at the regional level in 2013 and 2014; this chapter analyses the current state of play of climate-related development finance that was committed to the EECCA countries in 2013 and 2014 by bilateral and multilateral sources. It also provides some comparative analysis between the EECCA region and other regions of the world.
- Chapter 3. Key issues on access of the EECCA countries to scaled-up climate-related development finance; this part highlights issues around access of the EECCA

countries to international finance for climate actions, and outlines key analytical questions. Such questions aim to help countries better understand their levels of readiness and potential areas for improvement. This analysis is based on the literature review of existing readiness programmes implemented by several development co-operation agencies, international organisations and financial institutions.

- Chapter 4. Summaries of the country-level analyses of climate-related development finance in the 11 EECCA countries between 2011 and 2015; these summaries draw key findings from individual country-level reports on all the 11 EECCA countries. The 11 individual country-level reports are available on the OECD website: www.oecd.org/env/outreach/eap-tf.htm.

The full country-level reports include more detailed analysis of international development finance flows to support each EECCA country's climate actions. Each report also analyses the country's targets and priority sectors/areas in its climate actions based on the submitted INDC and other relevant policy documents. Finally, the country reports briefly outline enabling environments, such as climate-related policies and laws, as well as institutional arrangements and domestic funding entities or mechanisms that relate to investments in low-carbon and climate-resilient activities.

Policy recommendations

This part outlines key policy recommendations based on the analysis detailed in Chapters 2, 3 and 4, as well as the 11 individual country-level reports. The EECCA countries have already pursued climate actions through a range of policy reforms, capacity development and infrastructure investments, either by themselves or with development co-operation partners. They have been committed a significant amount of climate-related development finance (i.e. USD 3.3 billion to more than 300 projects per year during 2013-14 detailed in Chapter 2). Yet their INDCs and other policy documents suggest they still need significantly scaled-up finance from various sources to strive for low-carbon and climate-resilient development.

EECCA countries have great potential to be better prepared for accessing various kinds of climate finance sources. Such preparedness, or "readiness", to access climate finance consists of various elements such as stable climate goals and regulatory and policy frameworks; relevant institutional capacities; ability to develop bankable project pipelines; and monitoring and evaluation frameworks. Traditional public development finance sources such as Official Development Assistance (ODA) have struggled to support countries that have considerably limited policy frameworks and institutional arrangements. Mobilising private finance for such countries is even more challenging as the private sector tends to be less tolerant for such country risks and policy uncertainty (Kato, Ellis and Clapp, 2014).

Work under the OECD-hosted GREEN Action Programme intends to complement ongoing and planned activities in the region by international financial institutions, bilateral donors and multilateral development agencies. Development co-operation partners such as Deutsche Gesellschaft für Internationale Zusammenarbeit (GIZ), the KfW Development Bank, the United Nations Environment Programme (UNEP), the United Nations Development Programme (UNDP) and the World Resources Institute have been building capacity for climate finance readiness in some of the EECCA countries (see Chapter 3). The work under the GREEN Action Programme will also build on the OECD's experiences in working with the EECCA countries over the past 20 years (see OECD, 2016), as well as analytical activities under other committees of the organisation.

Ensure the implementation of INDCs and concrete actions towards climate targets

The INDCs of the EECCA countries and their other types of policy documents could benefit still further from concrete planning for effective and continuous implementation and enhanced coherence. The INDCs are expected to help build strong, durable and transparent foundations within the EECCA countries for their low-carbon, climate-resilient development. Apart from INDCs, these countries have a number of policies that directly or indirectly mitigate GHG emissions, and address adverse impacts of climate change. Such policies need to be effective to reduce climate risks, while being feasible and mutually reinforcing.

Many of these environment- and climate-related policy documents in the EECCA countries have faced challenges in implementation over the years. The documents were often prepared quickly, without a thorough and robust analysis by all relevant government officials, experts and other stakeholders within the country. This rushed process often undermines the translation of objectives into specific and time-bound targets; the creation of realistic financing and implementation strategies for their achievement; and the ability to design and implement investment projects that can bring actual results on the ground. Good capacity building requires the continuous involvement of government officials in the process of developing climate-related national and financing strategies and related regulations.

The EECCA countries should carefully examine how to ensure effective implementation of climate actions, and how to measure progress against their mitigation and adaptation targets. Some EECCA countries (e.g. Moldova, Georgia and Armenia) have already been developing national-level climate strategies that can serve as "action plans" for their NDCs, such as Low-Emission Development Strategies (LEDS) and National Adaptation Plans (NAPs), as well as country-specific national or sectoral plans on climate change, including Nationally Appropriate Mitigation Actions (NAMAs). Some of the results from the Technology Needs Assessments (TNAs) in many EECCA countries should be better used to improve actual uptake and market penetration of technologies needed for low-carbon, climate-resilient development.

Strengthen institutional capacity to better access climate finance

The landscape of international climate finance is rapidly evolving, but many countries struggle to seize emerging opportunities, including from relatively new sources such as the Green Climate Fund (GCF). Further work is needed for key institutions of the EECCA countries to better access and absorb international financial resources to implement climate actions. Given different needs and circumstances regarding access to climate finance, the countries may have varied priorities for capacity development. Possible capacities of such national institutions to be assessed include:

- understanding of national needs, priorities and strategies;
- ability to obtain updated information on various climate funds and other sources to support strategic decision-making on how to engage with those sources;
- familiarity with relevant institutions and stakeholders in the country and capability to operate a country co-ordination mechanism(s) and multi-stakeholder engagement;
- ability to programme and develop a pipeline of projects that can be financed;
- capability to facilitate project development and funding by national sources within the country;

- technical knowledge and skills to monitor and evaluate the effectiveness of the projects and programmes in light of national development and climate targets and in accordance with relevant guidelines of specific climate finance sources.

Some EECCA countries (e.g. Armenia, Georgia, Kyrgyzstan, Moldova and Tajikistan) have established entities to co-ordinate national climate actions. However, the extent to which such entities have sufficient capacities and resources, as well as appropriate authority to carry out their responsibilities, is often not clear. The questions outlined in Chapter 3 could help the EECCA countries self-assess gaps in such capacities and help their development co-operation partners to identify priorities for support.

Moreover, none of the EECCA countries has accessed climate finance to date through direct access modalities under the Adaptation Fund or the GCF (i.e. directly accessing the funding sources through the countries' own institutions rather than multilateral development banks, UN agencies, bilateral or multilateral development co-operation agencies). Any EECCA countries that wish to pursue a direct access modality under the GCF or the Adaptation Fund can usefully consider potential candidates for a national or regional implementing entity that can directly access certain types of climate finance sources. Some of the countries, such as Uzbekistan, have started to prepare for direct access. However, since finding such candidates and developing their capacities to be accredited is often time consuming, it is unlikely that direct access will become the favoured modality of EECCA countries in the short term. Thus, multilateral and bilateral providers of support will continue to play an important role in implementing climate-related projects.

Connect "dots" to enhance coherence between climate and non-climate policies

Better coherence among a range of climate and non-climate policies developed by the EECCA countries over the years can help the countries move efficiently towards low-carbon and climate-resilient economies, and mobilise further financial resources needed for their climate actions. To enhance such policy coherence, the EECCA countries first need to take stock of existing and planned policies in the country, and identify any misalignment. The countries then explore potential areas to improve coherence across such policies. Policy frameworks and measures to be examined could be broad and markedly different between countries, ranging from adaptation and mitigation, finance and investment promotion to competition, and other environmental and development issues.

Ultimately, such analysis should help the EECCA countries understand whether and how these policies can better interact and facilitate access to public and private climate finance for climate actions. Analytical methodologies could draw on the report on "Aligning Policies for a Low-carbon Economy" that highlights how policy misalignments in non-climate areas can impede efficient and cost-effective climate actions (OECD, 2015b). The analysis could also build on ongoing projects under the OECD Green Action Programme, such as on energy subsidy reforms and financial planning in the public sector in the region, as well as relevant work streams under other OECD committees such as the "Clean Energy Investment Policy Reviews" (OECD, 2015a).

Concrete action plans at a sectoral or sub-sectoral level, based on a comprehensive analysis of policy coherence, to achieve the countries' NDCs and calculate costs of necessary actions will help the countries better understand their need for support. It will also help them communicate such information to potential providers of finance, technologies and capacity building assistance. Such work could start with an in-depth discussion among relevant stakeholders on how to identify mitigation and adaptation options and to calculate the costs of their implementation, including a modelling exercise.

Improve transparency about climate finance flows

While this report aims to provide a clearer picture of climate-related development finance flows to the EECCA countries, there remains a significant information gap particularly with regard to financial flows from private sector and domestic sources, as well as from non-DAC member donors. For instance, some donors only report the aggregate amounts committed or disbursed, or do not report to the OECD Development Assistance Committee (DAC) Creditor Reporting System (CRS) at all. Kyrgyzstan had the smallest committed amount of climate-related development finance recorded in the OECD DAC CRS for the two years between 2013-14. However, substantial climate-related finance appears to flow from non-DAC member countries such as the Russian Federation (Government of Russian Federation, 2015). Moreover, as of 2014, the Chinese Export-Import Bank provided about 46% of Tajikistan's total external public debt as loans, although this was not limited to climate-related projects (World Bank, 2015). Further, about 80% of foreign direct investment flows to Azerbaijan were intended for the oil and gas sector in 2014; the OECD DAC CRS database shows the largest amounts of climate-related development finance in the country in 2013 and 2014 were committed to the waste management and road transport sectors. These examples illustrate the gap in information that prevents a comprehensive picture of climate-related financial flows into the country.

Transparency about the scale and type of climate finance provided, mobilised and received is important for domestic and international purposes. Better understanding of climate finance flows within the country, for example, could enable further discussion on how to improve the effectiveness of international public finance in mobilising broader capital resources. Such clearer information could help the EECCA countries strengthen trust with existing and potential providers of finance. More in-depth analysis can also help track the delivery of disbursed international climate finance to "end-users" (e.g. industry, households and sub-national governments) within a country. Such analysis can look into which domestic institutions, stakeholders, national funding entities and financial institutions are involved in the processes. The provisions of the Paris Agreement and Decision 1/CP.21 encourage, but do not require, the reporting of information on support received by developing countries.

A number of international institutions, including the OECD, are already working on issues around transparency in climate finance, and their recent advances and outcomes could inform future efforts by the EECCA countries to enhance it. Such efforts should also include identifying information and data gaps, as well as capacity-building needs in statistical authorities and bodies of the EECCA countries. All the above-mentioned analysis could lead to better evidence-based decision making in designing future climate targets and necessary policies, and help mobilise private sector investment and domestic financial resources.

Explore better use of national entities in mobilising climate finance

The experience of many developing countries shows the significant benefits from using in-country systems to co-ordinate and manage international and domestic climate-related finance. Such benefits include reducing duplication and transaction costs; enhancing a country's ownership over accessing and using financial resources; strengthening linkages between climate policies and the country's core planning and budgeting processes; and improving accountability and transparency.

An example of such in-country systems is national funding entities, which most EECCA countries have established with different sizes, structures and functions. A number of climate-related projects have received funding from such entities, some of which have been set-up with international support (e.g. the Energy Saving and Renewable Energy Fund in Armenia); the providers of support make sure these entities are designed in line with good international practices. The OECD has worked for a long time on strengthening the capacity of public finance institutions in the EECCA region. This experience could become the foundation to support both EECCA countries and their development co-operation partners in strengthening the capacity of such entities to better manage internal and external finance resources. For instance, such work could help the countries identify and select investment projects, monitor the effectiveness of their projects in light of the countries' climate and other development goals, and make use of financial resources more cost-efficient.

Further analytical work for the EECCA countries and their partners could include various elements. There include: reviewing, in the light of pursuing the targets under their INDCs, the rules and procedures relating to such national (funding) entities, which govern their decision-making process and management practices; examining the predictability and reliability of their sources of finance; analysing their expenditure plans and disbursement mechanisms; and understanding criteria for identifying and selecting investment projects for financing from the funding entity.

Note

1. Each Party is invited to communicate its nationally determined contribution (NDC) once the Party submits its respective instrument of ratification, accession or approval of the Paris Agreement. If the Party has communicated an intended nationally determined contribution (INDC), it can be considered to be the NDC unless the Party decides otherwise (Paragraph 22, Decision 1/CP.21).

References

Government of Russian Federation (2015), "Second biennial report common tabular format, submitted to the UNFCCC", *Second Biennial Reports*, http://unfccc.int/national_reports/biennial_reports_and_iar/submitted_biennial_reports/items/7550.php.

IEA (2016), *IEA CO_2 from Fuel Combustion Statistics* (database), http://dx.doi.org/10.1787/co2-data-en (accessed 24 October 2016).

IEA (2015), "Energy and climate change", *World Energy Outlook Special Report*, IEA/OECD Publishing, Paris, www.iea.org/publications/freepublications/publication/WEO2015SpecialReportonEnergyandClimateChange.pdf (accessed 24 October 2016).

Kato, T., J. Ellis and C. Clapp (2014), "The role of the 2015 agreement in mobilising climate finance", *OECD/IEA Climate Change Expert Group Papers*, No. 2014/07, OECD Publishing, Paris, http://dx.doi.org/10.1787/5js03h9ztlbr-en.

Melkonyan, A. (2015), "Climate change impact on water resources and crop production in Armenia", *Agricultural Water Management*, Vol. 161, Elsevier, Amsterdam, pp. 86-101.

Moore, R. (2015), "GEF support for intended nationally determined contributions (INDCs) & lessons learned", presentation at the Congo basin forest partnership (CBFP) meeting, Yaounde, 15-19 June 2015, http://pfbc-cbfp.org/rapports/items/stream-climat-fr.html?file=docs/news/RdP-15/Stream-Document-Climate/Climat-stream2-GEF%20Congo%20Basin_INDCs.pdf (accessed 24 October 2016).

OECD (2016), *Greening the Economy in Eastern Europe, the Caucasus and Central Asia*, June, OECD Publishing, Paris, http://issuu.com/oecd.publishing/docs/from_eap_to_green_action_programme_/1?e=3055080/36192948.

OECD (2015a), *Policy Guidance for Investment in Clean Energy Infrastructure: Expanding Access to Clean Energy for Green Growth and Development*, OECD Publishing, Paris, http://dx.doi.org/10.1787/9789264212664-en.

OECD (2015b), *Aligning Policies for a Low-carbon Economy*, OECD Publishing, Paris, http://dx.doi.org/10.1787/9789264233294-en.

UNFCCC (2016), *2014 Biennial Assessment and Overview of Climate Finance Flows Report*, http://unfccc.int/cooperation_and_support/financial_mechanism/standing_committee/items/8034.php (accessed 24 October 2016).

UNFCCC (2015), INDCs as communicated by Parties, http://www4.unfccc.int/submissions/INDC/Submission%20Pages/submissions.aspx (accessed 24 October 2016).

World Bank (2015), *The World Bank Group – Tajikistan Partnership Program Snapshot*, World Bank Group, Washington, DC, http://pubdocs.worldbank.org/pubdocs/public-doc/2015/10/645741444794465533/Tajikistan-Snapshot.pdf.

[illegible] (2016) [illegible] 2015 [illegible] OECD Publishing, Paris [illegible].

[illegible] (accessed 24 October 2016).

OECD (2016), [illegible], OECD Publishing, Paris, [illegible].

OECD (2016), [illegible] Clean Energy for Green Growth and Development, OECD Publishing, Paris, [illegible].

OECD (20[illegible]), [illegible], OECD Publishing, Paris, [illegible].

[illegible] 2016.

World Bank (2016), [illegible].

Chapter 2

Landscape of climate-related development finance at regional level

This chapter examines the climate-related development finance committed to the Eastern Europe, the Caucasus and Central Asia (EECCA) region in the two years between 2013-14. Multilateral and bilateral sources committed about USD 3.3 billion of climate-related development finance per year to the region during this period. The analysis also illustrates a great potential to mainstream climate concerns into development finance, especially for agriculture, forestry, industry, mining, water and transport sectors. It is primarily based on datasets from the OECD Creditor Reporting System Aid Activities Database. The Reader's Guide offers further information on data sources and the analytical framework.

Background and analytical methodologies

This chapter examines the climate-related development finance flow committed to the countries of Eastern Europe, the Caucasus and Central Asia (EECCA) in 2013 and 2014 by bilateral and multilateral sources. Data on the flows of climate-related development finance were obtained from the Creditor Reporting System (CRS) Aid Activities Database,[1] managed by the OECD Development Assistance Committee (DAC) (OECD, 2016a). This database allows for an approximate count of climate-related development finance flows that target climate mitigation and adaptation objectives of the Rio Conventions (including the UNFCCC).

The term "climate-related development finance" is used to mean finance committed by bilateral and multilateral sources to activities in developing countries that target climate mitigation or adaptation as either their principle objective or significant objective. On the DAC CRS, an activity is classified as related to climate change mitigation if it helps stabilise greenhouse gas (GHG) concentrations in the atmosphere by promoting efforts to reduce or limit GHG emissions or to enhance GHG sequestration. An activity is classified as related to climate change adaptation if it intends to reduce the vulnerability of human or natural systems to the current and expected impacts of climate change by maintaining or increasing resilience; by increasing ability to adapt to, or absorb, climate change stresses, shocks and variability; and/or by helping reduce exposure to them (OECD, 2016a).

Bilateral sources that report to the DAC CRS are DAC members, while the multilateral sources include multilateral development banks (MDBs) and international climate funds. The DAC CRS also tracks flows from some South-South co-operation sources. This database does not include financial flows at the activity-level from some non-DAC member donors such as the People's Republic of China (hereafter "China") and the Russian Federation (hereafter "Russia") or private sector sources, which are presumably providing a significant amount of finance to the EECCA region. This study presents only a snapshot for the years 2013 and 2014 due to limited data availability. The DAC CRS records the face values of activities on the dates when recipients sign grant or loan agreements (i.e. commitment of funds, but not disbursement). Therefore, there may be gaps between results from the DAC CRS and recipient countries' own statistics on external climate-related development finance, especially when observed over a longer period.

Out of 11 EECCA countries, Belarus, Kazakhstan and Ukraine are Annex I Parties to the United Nations Framework Convention on Climate Change (UNFCCC), while the others are non-Annex I Parties. Nonetheless, all 11 countries are eligible to receive official development assistance (ODA) from OECD DAC member countries. They also have access to some of the funding sources under the Financial Mechanism of the Convention[2] (e.g. Global Environment Facility, or the GEF). Exceptions include the Least Developed Countries Fund operated by the GEF since none of the EECCA countries is categorised as a Least Developed Country under the UN. While the data analysed in this chapter is on climate-related development finance committed in 2013 and 2014, the Board of the Green Climate Fund approved two project proposals in the EECCA region at its 13th meeting in June 2016 (one in Armenia and the other in Tajikistan and Uzbekistan).

Chapter 4 summarises the country-level analysis of the 11 EECCA countries. It outlines information on climate-related development finance and sectors where such finance is directed, as well as policies directly or indirectly relating to mobilisation of finance for climate actions.

Climate-related development finance flows to EECCA and other developing countries

In 2013-14, approximately USD 3.3 billion of climate-related development finance was committed per year to the EECCA region with a considerable difference in size that each country has received, or expects to receive (Figure 2.1). This amount covers both finance for adaptation and mitigation, while excluding overlapping finance for both mitigation and adaptation. It accounts for 7.1% of the total climate-related development finance committed globally in 2013 and 2014 (i.e. USD 47.3 billion per year on two-year average). More specifically, climate-related development finance committed to mitigation in the EECCA is 8.3% of that for all the recipients globally, while adaptation finance accounts for 3.5% of the global total. Further, bilateral donors and multilateral financial institutions committed to support more than 660 projects in these countries in those two years.

In 2013 and 2014, climate-related development finance for mitigation in the region (USD 3.0 billion per year) is about five times larger than that for adaptation (USD 0.62 billion per year). Although the Paris Agreement has stressed the importance of balancing financial resources between adaptation and mitigation (UNFCCC, 2015), the imbalance is a tendency observed globally. However, the imbalance is more pronounced in the EECCA countries where mitigation accounts for 81% of funds, adaptation for 11% and both together for 8%; the global average is 63% for mitigation, 25% for adaptation and 12% for both. The larger imbalance may be partly due to the lower level of policy development on adaptation in many of the EECCA countries (see also Chapter 3), but also to the high investment needs for mitigation in the region's energy sector. In addition, none of the 11 EECCA countries falls under the category Least Developed Countries or Small Island Developing States, which often have greater adaptation needs. Finally, it is generally more difficult to track finance for adaptation since it tends to be embedded into broader development projects and/or business operations, and thus, may be underestimated (CPI and OECD, 2015).

Figure 2.1. **Annual climate-related development finance flows for mitigation and adaptation: Comparison between EECCA countries and the global total for all recipients (Average 2013-14)**

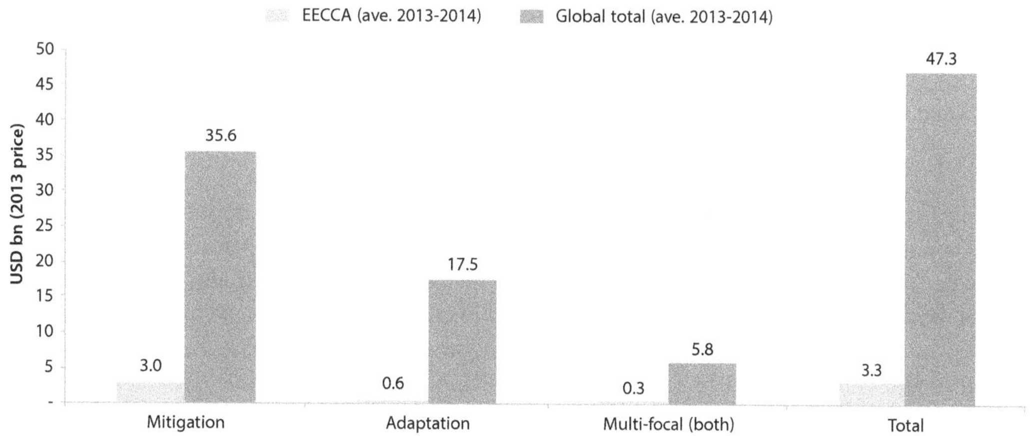

Note: Total climate-related development finance = Mitigation + Adaptation – Overlap (both).

Source: OECD (2016a).

Comparison between the EECCA and other regions

At the regional level, the amount of climate finance committed to the EECCA countries in 2013 and 2014 is similar to the amount committed to Southeast Asia, Southeast Europe, North and Central America, South America and North of Sahara regions (Figure 2.2). Largely due to its small population size, compared to other regions such as South of Sahara, South and Central Asia (excluding EECCA) and Southeast Asia, the EECCA region receives a relatively large amount of climate-related development finance per person. However, needs, socio-economic circumstances and populations differ significantly across countries. Thus, the volume of finance committed does not directly translate into how the needs of each country are being met, or into the difference in levels of readiness of each region to access international climate finance.

Figure 2.2. **Annual climate-related development finance flows by region (Total and per capita: average 2013-14)**

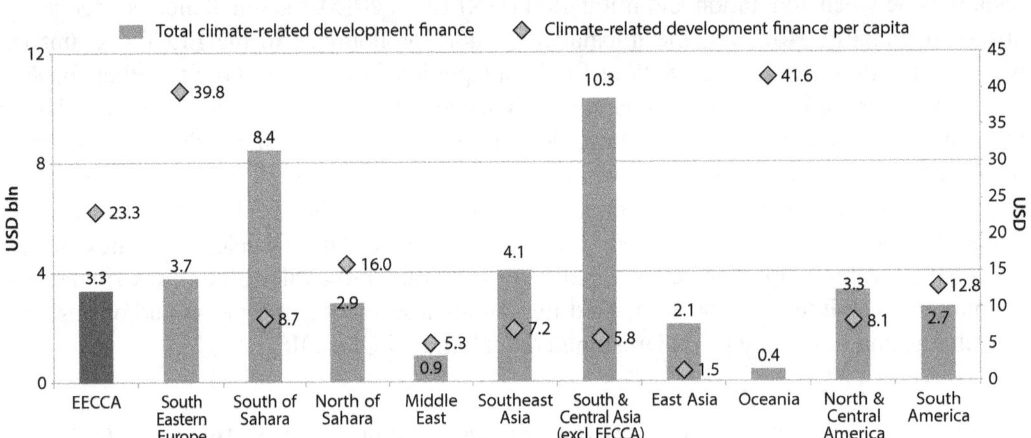

Notes: Bars represent climate-related development finance flows per year (USD billion, average between 2013 and 2014). Dots represent climate-related development finance flows per capita per year (USD per person, average between 2013 and 2014). The figure excludes some trans-boundary projects across countries/regions to avoid double counting.

Sources: Climate finance data adapted from OECD (2016a); GDP per capita purchasing power parity (PPP) from World Bank (2016).

Figure 2.3 plots the data on gross domestic product (GDP) per capita levels of those regions and the amounts of climate-related development finance committed. It illustrates the general trend that as economic development levels rise, committed amounts of climate-related development finance drop. Compared to the other regions in terms of GDP per capita levels, the EECCA region seems to have been committed a fair share of climate-related development finance. In 2013 and 2014, the EECCA region, as a whole, was committed a similar level of finance to levels committed to the North and Central America and the North of Sahara Africa regions. However, the EECCA region has much smaller population than these two regions. Thus, while the GDP per capita purchasing power parity (PPP) levels are slightly higher in these two regions than in the EECCA countries, the amounts of the finance per capita is much higher in the EECCA region as shown in Figure 2.2.

Figure 2.3. **Annual climate-related development finance and GDP per capita PPP by region**

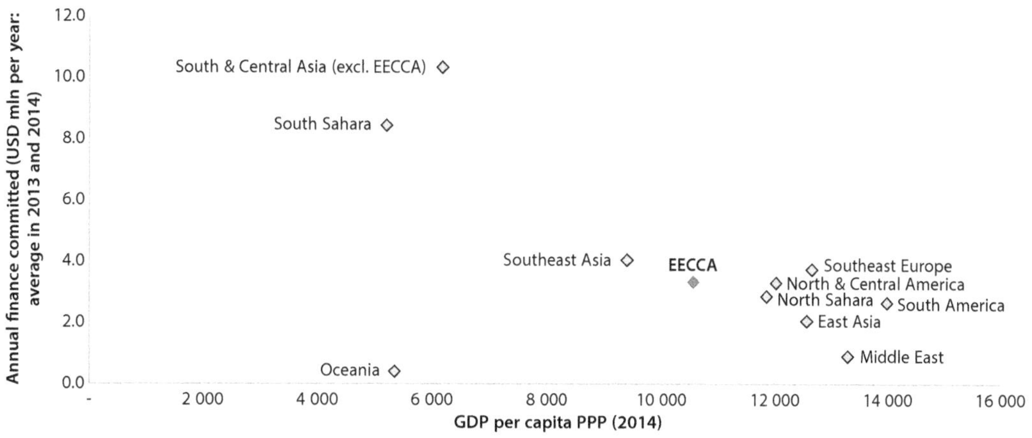

Annual finance committed (USD mln per year: average in 2013 and 2014)
12.0
10.0
8.0
6.0
4.0
2.0
0.0
South & Central Asia (excl. EECCA)
South Sahara
Southeast Asia
EECCA
Southeast Europe
North & Central America
North Sahara
South America
East Asia
Middle East
Oceania
-
2 000
4 000
6 000
8 000
10 000
12 000
14 000
16 000
GDP per capita PPP (2014)

Note: The data include only developing countries that are eligible to receive ODA.

Source: Climate finance data adapted from OECD (2016a); GDP per capita PPP from World Bank (2016).

Comparison across EECCA countries

For the two years between 2013-14, among the 11 EECCA countries, the largest amounts of climate finance in absolute values were committed to Ukraine and Uzbekistan, while Turkmenistan, Azerbaijan and Kyrgyzstan received relatively small volumes (Figure 2.4). Of all commitments in the EECCA countries in 2013-14, the six largest projects were directed to either Uzbekistan or Ukraine (i.e. those supported by Japan, the World Bank Group and the European Investment Bank). This has significantly affected the entire landscape of climate-related development finance committed to the region in 2013-14. Such projects include high-efficiency gas-fired power plants, district heating, energy efficiency, transport sector infrastructure investment and water resource management (for adaptation).

Figure 2.4. **Annual climate-related development finance flows by country in the EECCA region (Annual total and per capita: average 2013-14)**

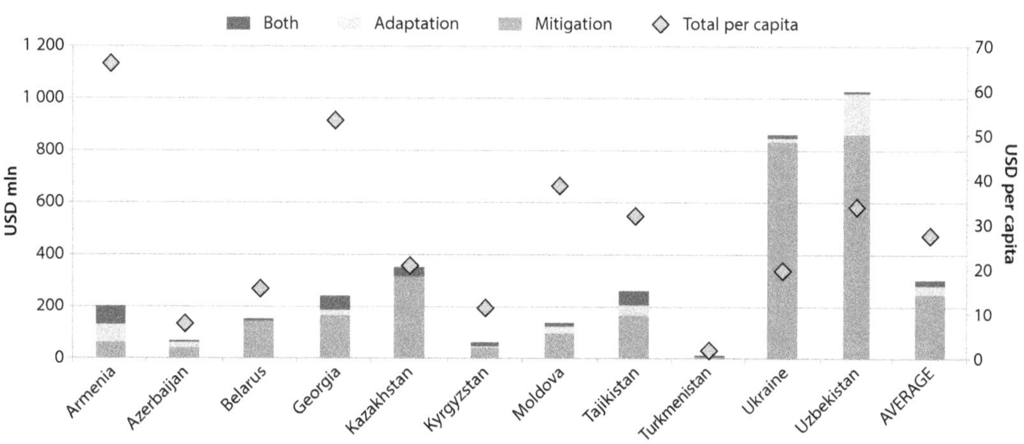

Source: Climate finance data adapted from OECD (2016a); GDP per capita PPP from World Bank (2016).

Among the resource-rich countries, Kazakhstan received a significantly larger amount of finance than Azerbaijan or Turkmenistan. Such differences reflect various factors, including size of population, needs for external finance, in-country policies and targets that can facilitate attracting international development finance, geopolitical circumstances and interests, and size of finance inflows from providers not included in the OECD DAC CRS database (e.g. China and Russia).

The less populated countries tend to receive considerably larger amounts of finance per capita than the average (i.e. USD 27 per capita annually on average) (Figure 2.4). The populations of these countries are 3.0 million (Armenia), 3.6 million (Moldova), 4.5 million (Georgia) and 8.4 million (Tajikistan). Uzbekistan with 45.6 million inhabitants also received a relatively large amount per capita finance, reflecting the large size of total climate-related development finance for the country. On the other hand, Azerbaijan, Kyrgyzstan and especially Turkmenistan received relatively small amounts of finance per capita. Moreover, they also received a smaller size of total climate-related development finance than the amount committed to most other EECCA countries. Azerbaijan and Turkmenistan have higher levels of domestic financial resources than Kyrgyzstan, which is one of the poorest countries in the region (See also Figure 2.5).

GDP per capita and climate-related development finance

Even within the group of countries with similar GDP per capita (PPP) levels, the amount of climate-related development finance committed to each EECCA country varies significantly (Figure 2.5). For instance, a relatively large amount of finance was committed to Uzbekistan and Ukraine (USD 1 billion per year and USD 860 million per year, respectively) compared to Moldova and Armenia, despite their similar GDP per capita (PPP) levels. This is partly attributed to a few large-scale infrastructure investments in energy, transport and water resource sectors in Uzbekistan and Ukraine in 2013 and 2014. These two countries have larger populations and are more industrialised than Moldova and Armenia, thus seem to have greater needs for infrastructure investments per se. On the other hand, at the per capital level, Armenia and Moldova were committed greater amounts than Uzbekistan and Ukraine.

Figure 2.5. **Annual climate-related development finance and GDP per capita PPP**

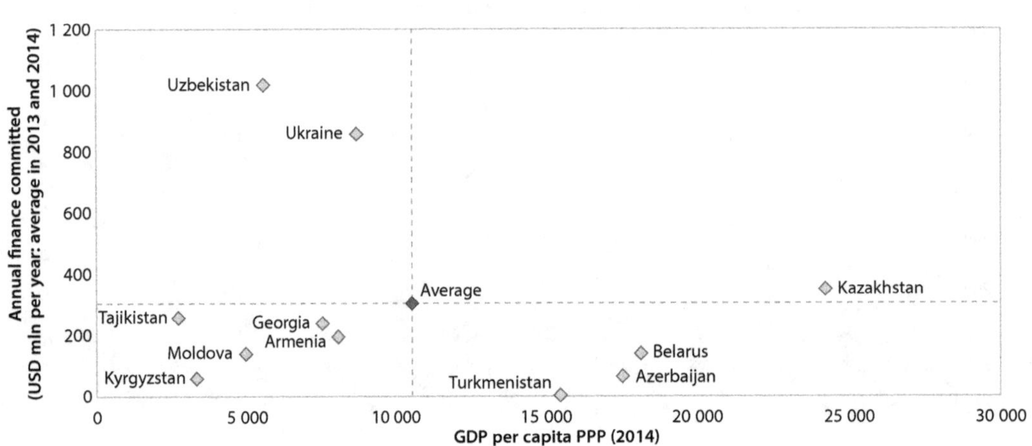

Source: Climate finance data adapted from OECD (2016a); GDP per capita PPP from World Bank (2016).

The amounts of climate-related development finance committed to Kazakhstan, Tajikistan and Georgia are relatively close to the average of all the 11 EECCA countries, whereas their per capita income levels significantly differ. Belarus, Azerbaijan and Turkmenistan have relatively high levels of GDP per capita (PPP). These countries may have been relying more on foreign direct investment and/or domestic financial resources, but some may not have a high level of demand for investment in climate-related projects. As a result, they receive relatively small amounts of climate-related development finance from international public sources. On the other hand, Kazakhstan, at a higher GDP per capita level than these three countries, was committed a larger level of climate-related development finance than the EECCA average.

International financial sources support a number of trans-boundary (or regional) projects, each of which covers more than one country in the region. Examples include projects or programmes on regional electricity transmission network improvement; flood management; better ecosystems and agriculture management for trans-boundary river basins; regional forest management; and international road transport infrastructure. Selected examples of such projects and programmes are listed below. The committed amounts to some regional projects were allocated to each participating country and recorded in the OECD DAC CRS, and are thus reflected in the figures by country. Others were unallocated and recorded as, for instance, "Europe regional".

- "Caucasus Electricity Transmission Network" project, supported by Germany, the European Union (EU) and the European Investment Bank (EIB), establishes better energy interconnections to enable energy exchange and transit between Armenia and Georgia, as well as other areas in the South Caucasus region.
- "Central Asia South Asia Electricity Transmission and Trade" project (CASA-1000) is a trans-boundary project implemented jointly by Kyrgyzstan, Tajikistan, Afghanistan and Pakistan and co-financed and supported by a range of multilateral and bilateral financiers.
- "Climate Change and Security in the Dniester River Basin" project, implemented by the UN Economic Commission for Europe and the Organization for Security and Co-operation in Europe, aims to increase the capacity of Moldova and Ukraine to implement adaptation measures and enhance trans-boundary co-operation between them.
- "Regional Forest Law Enforcement and Governance Phase 2 (FLEG II)", financed by the EU, aims to contribute to mitigation and adaptation through better forest management in Armenia, Azerbaijan, Belarus, Georgia, Moldova, Russia and Ukraine.
- "Green Logistics Program", supported by the Global Environment Facility (GEF) and the European Bank for Reconstruction and Development in the Black Sea and Mediterranean region, covers interventions to reduce the negative impacts of freight transport such as GHG emissions, air pollution, noise, vibration and accidents throughout the logistics supply chain in the region.
- The EU's Neighbourhood Investment Facility (NIF) aims at mobilising funding to finance capital-intensive infrastructure projects in the EU partner countries covered by the European Neighbourhood Policy (ENP) in sectors such as transport, energy, environment and social development. In the EECCA region, Armenia, Azerbaijan, Belarus,[3] Georgia, Moldova and Ukraine are eligible to receive the funding.

- "Climate Adaptation and Mitigation Program for Aral Sea Basin (CAMP4ASB)", supported by the World Bank Group, is a platform for sustained regional dialogue and knowledge sharing among all Central Asian countries (i.e. Tajikistan and Uzbekistan) on climate change across a broad range of sectors, including agriculture, disaster risk management, water and energy. The Green Climate Fund also approved funding for this programme in 2016 (hence it is not reflected in the statistical data examined in this report).
- "ClimaEast" project, funded by the EU, provides capacity building support to Eastern Partnership countries (i.e. Armenia, Azerbaijan, Belarus, Georgia, Moldova and Ukraine). The specific objective is to foster improved climate change policies, strategies and economic instruments.

Sectoral-level analysis of climate-related development finance

Climate mitigation actions in the energy generation and supply sector (e.g. generation and distribution of electricity and heat) were committed the largest volume of climate-related development finance in 2013 and 2014 (Figure 2.6). This reflects the EECCA region's large financial needs to replace or rehabilitate aged and inefficient power plants and transmission infrastructure (IEA, 2015). The financing for energy generation and supply sector amounts to more than 40% of the total climate-related development finance committed to the EECCA countries during that period. This figure is significantly larger than the total committed finance to this sector at the global level (27%).

Figure 2.6. **Annual climate-related development finance by sector in EECCA countries (USD million in 2013)**

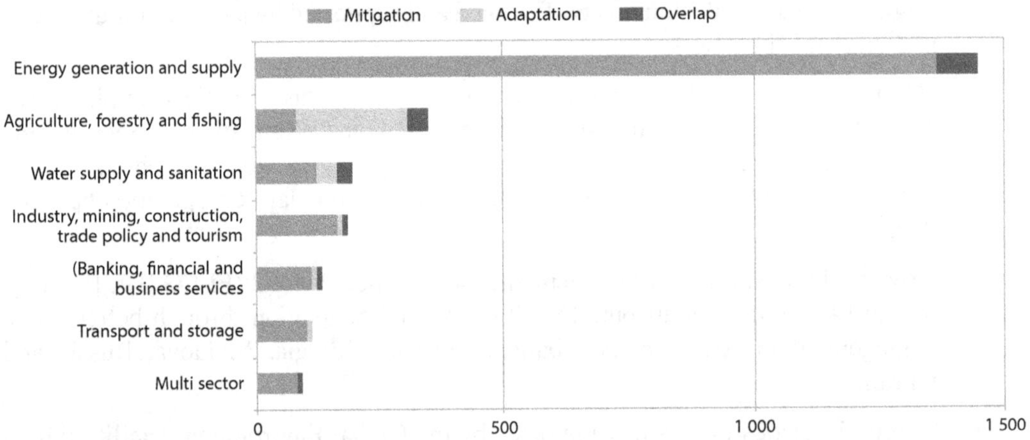

Source: Climate finance data adapted from OECD (2016a).

Given that energy-efficient technology has low market penetration in many EECCA countries, there still appears to be significant potential to invest in low-carbon energy infrastructure and equipment (EBRD, 2014; IEA, 2015). Especially for countries at lower levels of economic development, such as Kyrgyzstan and Tajikistan, the development of the energy sector is crucial in eradicating poverty and pursuing sustainable development. Further investment in energy infrastructure is also essential to enhance energy security in most of the EECCA countries, especially those that lack large natural resource reserves. Energy efficiency activities in the housing sector (e.g. better insulation) can help significantly reduce energy-related poverty. For instance, the low energy-efficiency of buildings and equipment used in houses and apartments contributes to the high amounts that residents spend each month on energy. In Tajikistan, for example, energy represents 14-25% of total household expenses during winter (GERES, 2014). Hydropower development may have both mitigation and adaptation aspects, especially in countries that largely rely on hydropower such as Armenia, Georgia, Kyrgyzstan and Tajikistan. This is because electricity generation is likely to be affected by a decrease in precipitation, more inconsistent rainfalls and a decrease in spring run-off from glaciers and snow caused by climate change.

A significant portion of climate-related development finance in the region was also committed to the agriculture, forestry and fishing sectors, as well as to water supply and sanitation, which are especially important for adaptation. About 10% of climate-related development finance was committed to agriculture, forestry and fishing, which is almost the same share as the global level. A range of actions in agriculture can contribute to poverty alleviation and low-carbon and climate-resilient development in the rural areas in some EECCA countries. The region has widely implemented projects on irrigation and drainage management. In the water sector, upgrading assets that are operating beyond their planned lifetime is a primary issue, even in countries with relatively developed water and wastewater infrastructure such as Ukraine (World Bank, 2014). Another issue is that development plans and investments regarding water infrastructure often do not seem to consider the long-term implications of climate change in many developing countries, including some EECCA countries.

Implementing climate change adaptation actions also has the potential to generate co-benefits for improving public health. This can be seen in a number of projects in improved water supply and sanitation, agriculture and food production, and natural disaster risk management (See the country reports for more details). Affordable agriculture insurance (e.g. crop insurance) could play an important role in helping farmers better financially prepare for climate risk by easing their economic losses when natural disasters occur. Yet there has not been significant up-take of such insurance products in the EECCA countries.

Some countries whose INDCs explicitly mention agriculture and forestry as priority sectors (e.g. Belarus, Georgia and Kyrgyzstan) saw considerably smaller commitments of adaptation finance than of mitigation finance for other sectors, such as energy, in 2013 and 2014. For instance, the size of the mitigation and adaptation finance committed to the agriculture and forestry sectors in Belarus in 2013 and 2014 was almost zero. In addition, while Georgia's INDC estimated total annual adaptation costs at USD 150-200 million between 2021-30 (Government of Georgia, 2015), climate-related development finance for adaptation was USD 19 million per year in 2013 and 2014. Domestic sources in these countries can often (co-)finance investments in these sectors; it would be useful to further explore ways to finance climate actions in these sectors.

In the banking and financial sector, a number of local banks in the region supported investments in energy efficiency and renewable energies projects, using credit lines from international sources of finance. Such international sources include the Asian Development Bank, European Bank for Reconstruction and Development (EBRD), the European Investment Bank (EIB), EU, International Finance Corporation and the KfW Development Bank, among others. Such projects using environmental credit lines often provide technical assistance to the local financial institutions as well, with support of donors such as the EU and the Climate Investment Funds in the form of grants or concessional loans. Such support aims to strengthen the capacities and knowledge base of local institutions regarding technologies and financial instruments. For further details, see each country report and OECD (2016c).

Mainstreaming of environment and climate into development finance

Figure 2.7 shows the share of development finance devoted to climate from the same multilateral and bilateral sources in key sectors in the EECCA. It is notable that 69% of development finance for the energy sector is climate-related, and particularly that finance for mitigation is higher than the global average (60%). There seems to be room for projects that aim to increase the resilience of the energy sector to climate impacts through adaptation, particularly for those especially vulnerable to climate change such as Tajikistan and Kyrgyzstan. A larger share of climate-related development finance in agriculture was committed to adaptation, but the total of 26% (including both mitigation and adaptation) is much lower than the global average (40% for climate-related development finance).

Reducing GHG emissions and vulnerability to climate change in the transport sector can also be further emphasised, as little development finance is directed to low-carbon or climate-resilient transport systems (i.e. 5% in EECCA vs. 31% in the world). For the water sector, the share of climate-related finance in the total development finance in the EECCA countries is the same as the global average (i.e. 37%). However, the level of committed finance to mitigation in the sector is much higher (i.e. 21% in EECCA vs. 3% in the world), reflecting the need to rehabilitate old, inefficient infrastructure in this sector. For further analysis of mainstreaming of development finance into climate action at the global level, see OECD (2015b.)

Figure 2.7. **Climate-related development finance as a share of total bilateral and multilateral development finance in the key sectors in EECCA**

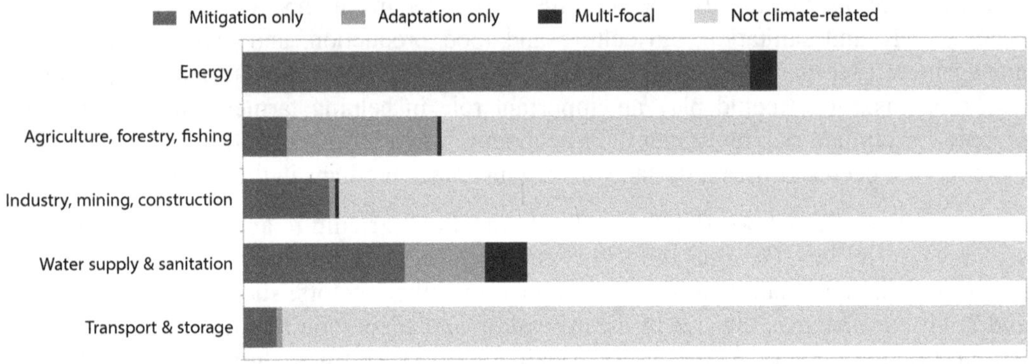

Source: OECD (2016a).

Delivery channels

MDBs channelled a greater level of climate-related development finance to the EECCA region than bilateral donors during 2013-14 with 61.7% (multilateral) and 29.6% (bilateral) of the total, respectively (Figure 2.8). This contrasts with the average of all recipient countries around the world where a significantly larger share of climate finance is delivered bilaterally (51.0%) than multilaterally (37.5%). Nevertheless, most ODA-recipient EECCA countries are either lower middle-income countries (USD 1 046-4 125 GNI per capita) or upper middle income countries (USD 4 126-12 745 GNI per capita). In such countries, loans from multilateral sources tend to play a greater role in financing climate and broader development activities than in least developed countries that often rely more on grants. The practice of loans from public sources also aims to avoid market distortion and crowding out of private-sector financing.

The major channels to deliver climate-related development finance differ across the EECCA countries. In general, as economic development levels increase and/or the domestic financial systems mature, finance tends to be delivered through multilateral channels. However, a range of other factors also affect delivery of climate finance, such as geopolitical interests and historical relationship with finance providers. Given the data present a snapshot of 2013 and 2014, a few large-scale projects committed during the period could distort the overall picture. For instance, Armenia (e.g. infrastructure in energy, water and agriculture sectors supported by Germany) and Uzbekistan (e.g. large-scale power plants supported by Japan) receive a large portion of finance through bilateral channels. Turkmenistan receives the smallest amount of climate-related development finance among the EECCA countries, which is delivered through the Global Environment Facility (for energy and agriculture sectors). Such large-scale projects have affected the compositions of support committed to these countries in 2013 and 2014 (see Figure 2.8 for data by country).

Figure 2.8. **Channels to deliver climate-related development finance**

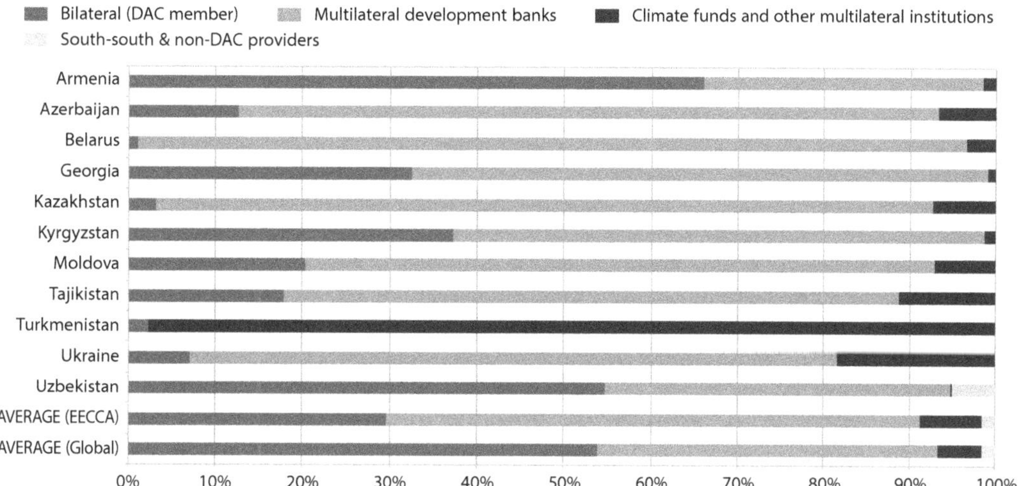

Source: OECD (2016a).

Among multilateral providers of finance, the European Bank for Reconstruction and Development (EBRD), the World Bank Group (i.e. IBRD, IDA and IFC), the Asian Development Bank (ADB) and the European Investment Bank (EIB) have committed a significant amount of climate-related development finance (Figure 2.9.) to a variety of sectors. Multilateral funds dedicated to climate action are also active in many EECCA countries. These include the Climate Investment Funds (e.g. district heating energy efficiency and renewable energy plants in Ukraine, adaptation programmes for rural areas and energy sector in Tajikistan) and funds operated by the Global Environment Facility (e.g. a variety of capacity building programmes in addition to capital investments).

Among bilateral donors, Japan committed the largest amount of climate-related development finance in 2013 and 2014, largely due to the large-scale gas-fired power plants in Turakurgan and Navoi in Uzbekistan, among others. Germany, the EU and Switzerland, also major providers of climate-related development finance, have committed to finance a relatively large number of projects, notably on adaptation for water resource management, agriculture and forestry, but also on mitigation. Bilateral donors also provide climate-related development finance through MDBs and climate funds, whose amount is included in the figures on the multilateral channels.

Figure 2.9. **Major providers of climate-related develpement finance to EECCA countries**

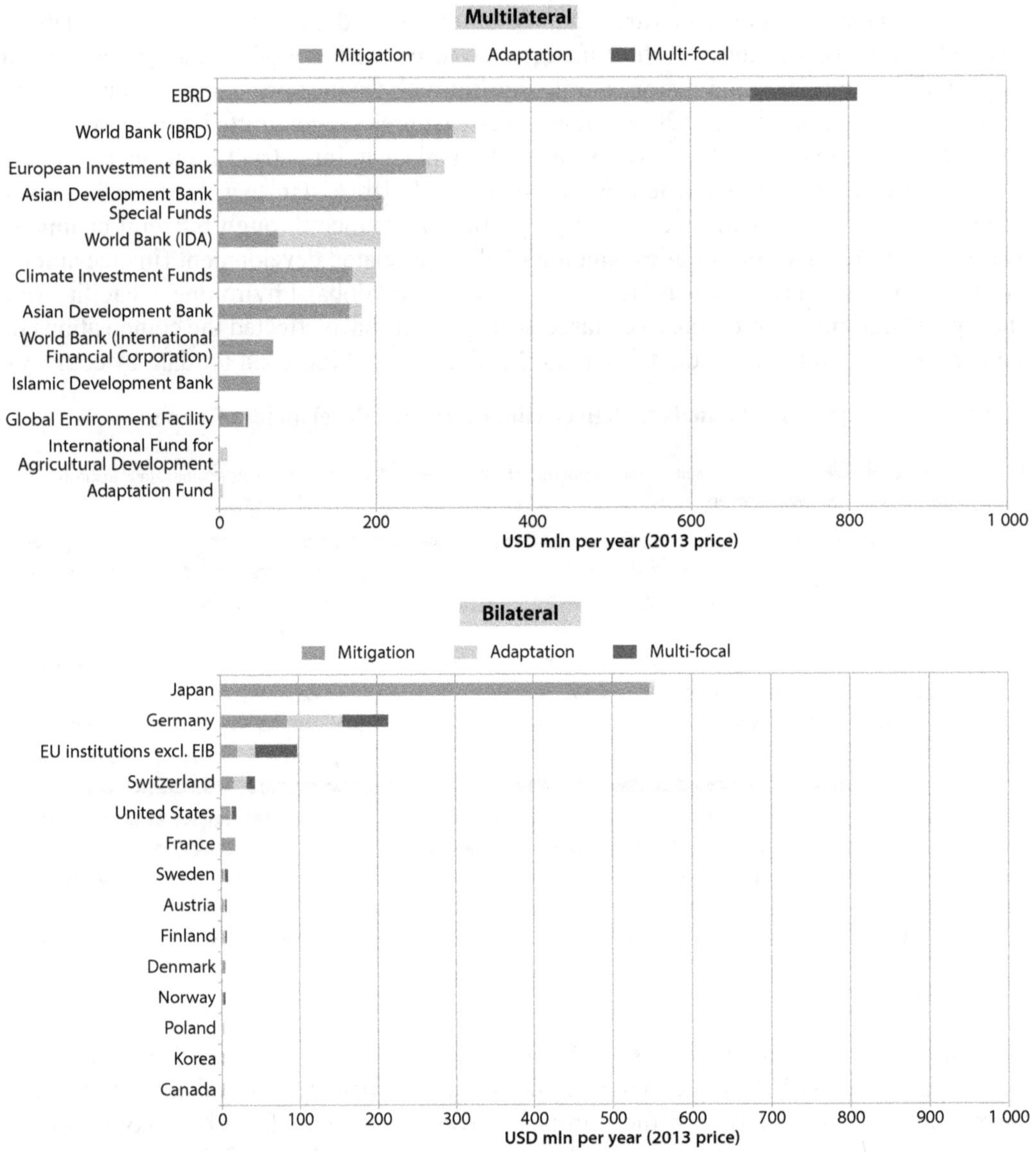

Notes: Financing by the Climate Investment Funds is channelled through the Clean Technology Fund (CTF) or the Pilot Program for Climate Resilience (PPCR).

Source: OECD (2016a).

In the last decade, the number of international dedicated climate funds has been increasing rapidly. The channels to deliver climate finance have also become more diverse than ever. Table 2.1. lists examples of such international climate funds and facilities, and indicates whether individual EECCA countries have accessed each of these funding sources. Many EECCA countries have already benefited from international climate funds that are either established as standalone funds or financed by bilateral and multilateral financiers.

Table 2.1. **Selected examples of dedicated climate funds/facilities (as of August 2016)**

Names of funds/facilities	ARM	AZE	BLR	GEO	KAZ	KYR	MDA	TJK	TKM	UKR	UZB
Green Climate Fund	X							X			X
Global Environment Facility (Trust Fund: Climate focus)	X	X	X	X	X	X	X	X	X	X	X
Special Climate Change Fund (operated by the GEF)		X		X	X	X	X	X	X		X
Adaptation Fund under UNFCCC				X					X		X
Germany's International Climate Initiative (IKI)	X	X	X	X	X	X	X	X	X	X	X
Eastern Europe Energy Efficiency and Environmental Partnership (E5P)	X	X	X	X			X			X	
EU Neighbourhood Investment Facility (NIF)	X	X	X	X			X			X	
Clean Technology Fund (of the Climate Investment Funds)					X					X	
Pilot Program for Climate Resilience (of the Climate Investment Funds)						X		X			
Scaling-Up Renewable Energy Program for Low-Income Countries (of the Climate Investment Funds)	X										
EBRD's Sustainable Energy Finance Facilities	X		X	X	X	X	X	*		X	
Global Facility for Disaster Reduction and Recovery				X		X	X	X	X		X
Global Energy Efficiency and Renewable Energy Fund (GEEREF)				X							
Climate Technology Initiative (CTI) Private Financing Advisory Network (PFAN)	X	X		X			X			X	
International Climate Fund (UK)					X					X	
Nordic Environment Finance Corporation (NEFCO)'s funds										X	
Partnership for Market Readiness										X	
Adaptation for Smallholder Agriculture Programme						X					
Investment Facility for Central Asia (IFCA)					X	X		X	X		X
Green for Growth Fund in Southeast Europe	X	X		X			X			X	
Global Climate Partnership Fund										X	

Note: The EBRD launched "the Climate Resilience Financing Facility" for Tajikistan in collaboration with the CIFs.

Sources: CIF (n.d.a., n.d.b., n.d.c.); CTI PFAN (n.d.); EBRD (n.d.a., n.d.b.) ; EIB (n.d.), GFDRR (n.d.); Global Climate Partnership Fund (n.d.); Green for Growth Fund (n.d.); IFAD (n.d.); Nordic Environment Finance Corporation (n.d.); European Commission (2015a, 2015b); International Climate Fund (2015); OECD (2015a); Adaptation Fund (2016); GCF (2016); GEF (2016); German Federal Ministry for the Environment, Nature Conservation, Building and Nuclear Safety (2016); Partnership for Market Readiness (2016).

The Financial Mechanism of the UNFCCC has two operating entities, namely the Green Climate Fund (GCF) and the Global Environment Facility (GEF). The GEF is also an operating entity for other multilateral environmental treaties such as the Convention on Biological Diversity and the Convention to Combat Desertification. All the EECCA countries have accessed climate-related development finance from the GEF. No data on finance approved by the GCF is included in the analysis in this chapter (as the database used here covers the period 2013-14), while the GCF Board has approved two projects from the EECCA countries as of 1 July 2016. These are the project on de-risking and scaling-up investment in energy-efficient building retrofits in Armenia with UNDP (GCF funding: USD 20 million); and the project to support the World Bank's climate adaptation and mitigation programme for the Aral Sea Basin in Tajikistan and Uzbekistan (GCF funding: USD 19 million). Both projects obtain the funding through international access modalities through UNDP and the World Bank, respectively.

Also, all the EU's Eastern Neighbourhood countries (i.e. Armenia, Azerbaijan, Belarus, Georgia, Moldova and Ukraine) have access to funding from the EU's Neighbourhood Investment Facility, and the Eastern Europe Energy Efficiency and Environmental Partnership (E5P). Germany's International Climate Initiative and EBRD's Sustainable Energy Finance Facilities have supported a large number of projects in EECCA countries.

Financial instruments

Various financial instruments, such as grants, concessional loans, commercial loans and equities, are used to deliver climate-related development finance to the EECCA countries (Figure 2.10). Non-concessional loans were the form most used to deliver climate finance to the EECCA countries in 2013 and 2014 and provided by MDBs. Bilateral channels are the main providers of grants and concessional loans through technical assistance, capital investments and co-financing to projects supported by non-concessional loans from MDBs. Bilateral donors also provide financing through multilateral donors (e.g. voluntary contributions to climate funds). Some of the MDBs and climate funds (e.g. the International Development Association of the World Bank Group and the Global Environment Facility) also provide grants and concessional finance to the EECCA region.

Figure 2.10. **Financial instruments and channels used to deliver finance to the EECCA in 2013-14**

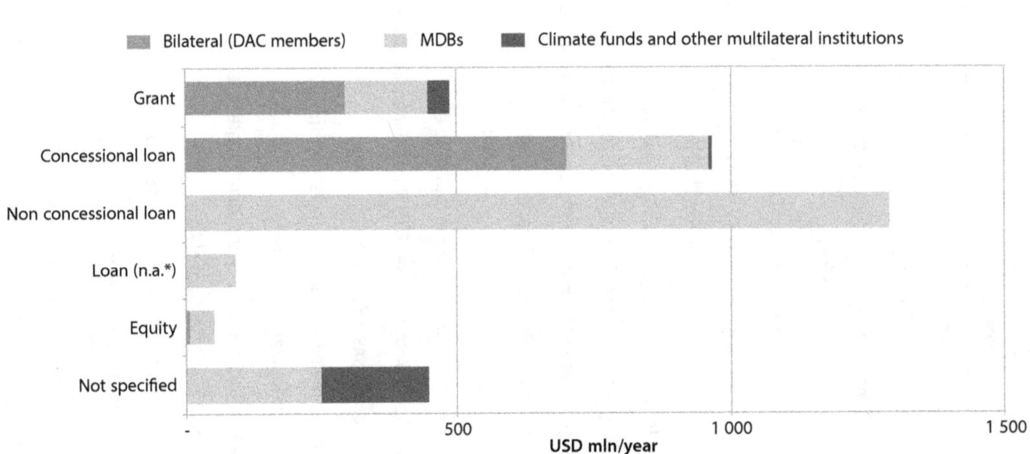

Source: OECD (2016a).

Grants are often used to support technical assistance, which plays a significant role in directly and indirectly mobilising finance for climate actions. For instance, technical assistance contributes to filling knowledge gaps and raising countries' awareness of climate risks and the importance of addressing them. This leads to increasing the capacity of stakeholders to design and implement policies to tackle climate risks. Technical assistance also helps reduce investment risks by, for instance, providing technology-related data on specific measures, as well as climate- and weather-related information. This can help potential project proponents and investors to develop project plans and make financial decisions (CPI and OECD, 2015).

Technical assistance in the EECCA countries includes capacity building projects such as development of Nationally Appropriate Mitigation Actions, preparation of the National Communications to the UNFCCC and the INDCs, and more general policy development support and capacity building activities for staff members and other stakeholders within the countries. The EU, for example, provides a large-scale grant financing to EBRD's Sustainable Energy Financing Facilities through the EU Neighbourhood Investment Facility to help local financial institutions develop their capacities in financing mitigation projects.

Financial instruments used for climate actions differ greatly among the EECCA countries (Figure 2.11) because the choice of such instruments depends on, for instance, project types, technologies employed, in-country financial systems and country-specific risks. Whereas most countries predominantly use loans (e.g. concessional loan financing for power plants, irrigation systems and transport infrastructure), Tajikistan and Kyrgyzstan receive a relatively large share of grant finance, reflecting their lower income levels. Examples of grant finance include the ADB's grant for Golovnaya hydropower plant (Tajikistan) and adaptation projects in forestry and agriculture sectors (Kyrgyzstan) mainly supported by Germany. The share of equity financing, which is small throughout the region, includes the European Investment Bank and KfW Entwicklungsbank in the Green for Growth Fund II (e.g. in Armenia, Ukraine and Georgia), and a hydropower project supported by the ADB (in Armenia).

Figure 2.11. **Financial instruments used in the region**

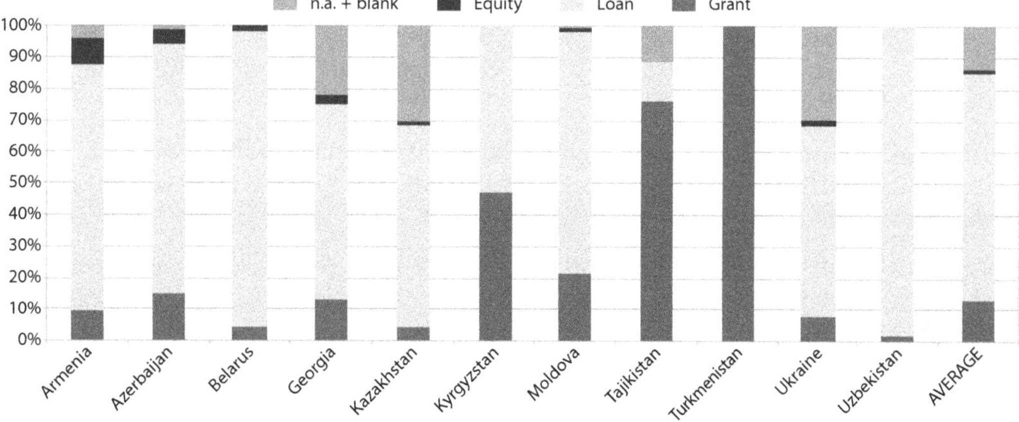

Source: OECD (2016a).

Climate-related projects and programmes are often financed by the combination of different types of financial instruments. For instance, the Clean Technology Fund has provided concessional loans to a range of projects in the EECCA countries for demonstration, deployment and transfer of low-carbon technologies. Most of the concessional loans are co-financing for projects supported by the EBRD using non-concessional loans.

References

ADB (2013), "Kyrgyz Republic: Water Supply and Sanitation Strategy", webpage, www.adb.org/projects/46350-001/main (accessed 24 October 2016).

Adaptation Fund (2016), "Projects Map View", webpage, www.adaptation-fund.org/projects-programmes/project-information/projects-map-view/ (accessed 24 October 2016).

CIF (n.d.a), Clean Technology Fund website, www-cif.climateinvestmentfunds.org/fund/clean-technology-fund (accessed 24 October 2016).

CIF (n.d.b), Pilot Program for Climate Resilience website, www-cif.climateinvestmentfunds.org/fund/pilot-program-climate-resilience (accessed 24 October 2016).

CIF (n.d.c), Scaling Up Renewable Energy Program website, www-cif.climateinvestmentfunds.org/fund/scaling-renewable-energy-program (accessed 24 October 2016).

CPI and OECD (2015), *Estimating Mobilized Private Finance for Adaptation: Exploring Data and Methods*, Climate Policy Initiative, London and OECD Publishing, Paris, http://climatepolicyinitiative.org/publication/estimating-mobilized-private-finance-for-adaptation-exploring-data-and-methods/.

CTI PFAN (n.d.), "CTI Private Financing Advisory Network", webpage, www.cti-pfan.net/region/?rp=cis-central-asia (accessed 24 October 2016).

EBRD (2014), "CEEP – Demirbank Azerbaijan", webpage, www.ebrd.com/work-with-us/projects/psd/ceep---demirbank-azerbaijan.html (accessed 24 October 2016).

EBRD (n.d.a), A flagship multi-donor fund to support energy efficiency and environmental investments in the Eastern Partnership region, http://e5p.eu/wp-content/uploads/2014/12/E5P%20Brochure-English.pdf (accessed 2 November 2016).

EBRD (n.d.b) "Sustainable Energy Financing Facilities", webpage, www.ebrd.com/what-we-do/sectors-and-topics/sustainable-resources/seffs.html (accessed 24 October 2016).

EIB (n.d.), "Global Energy Efficiency and Renewable Energy Fund", webpage, www.geeref.com/.

European Commission (2015a), "Neighbourhood Investment Facility", webpage, www.ec.europa.eu/enlargement/neighbourhood/neighbourhood-wide/neighbourhood-investment-facility/index_en.htm.

European Commission (2015b), *Investment Facility for Central Asia, Asia Investment Facility and Investment Facility for the Pacific Operational Report 2013 and 2014*, www.ec.europa.eu/europeaid/investment-facility-central-asia-ifca-asia-investment-facility-aif-investment-facility-pacific-ifp_en.

GCF (2016), "Green Climate Fund Portfolio", webpage, www.greenclimate.fund/ventures/portfolio (accessed 1 September 2016).

GEF (2016), "GEF Projects", www.thegef.org/projects (accessed 24 October 2016).

GERES (2014), "Energy Efficiency for Improved Living Conditions in Tajikistan", webpage, www.geres.eu/images/fiches/Project-PIGHOT-Tajikistan-EN.pdf (accessed 24 October 2016).

German Federal Ministry for the Environment, Nature Conservation, Building and Nuclear Safety (2016), "International Climate Initiative (IKI)", webpage, www.international-climate-initiative.com/en/nc/projects/projects/ (accessed 24 October 2016).

GFDRR (n.d.) "Global Facility for Disaster Reduction and Recovery: Europe and Central Asia", webpage, www.gfdrr.org/region/europe-and-central-asia (accessed 24 October 2016).

Global Climate Partnership Fund (n.d.), Portfolio website, www.gcpf.lu/portfolio.html (accessed 24 October 2016).

Green for Growth Fund (n.d.), Project Portfolio website, www.gcpf.lu/portfolio.html (accessed 24 October 2016).

Government of Georgia (2015), *Georgia's Intended Nationally Determined Contribution: Submission to the UNFCCC*, Ministry of Environment and Natural Resources Protection, Tbilisi, Georgia, www4.unfccc.int/submissions/INDC/Published%20Documents/Georgia/1/INDC_of_Georgia.pdf.

IEA (2015), *Energy Policies Beyond IEA Countries: Caspian and Black Sea Regions 2015*, IEA/OECD Publishing, Paris, http://dx.doi.org/10.1787/9789264228719-en.

IFAD (n.d.), "Adaptation for Smallholder Agriculture Programme (ASAP)", webpage, www.ifad.org/en/topic/asap/tags/climate_change/2782790 (accessed 24 October 2016).

International Climate Fund (2015), "Our Portfolio across the World", webpage, www.gov.uk/government/publications/international-climate-fund/international-climate-fund#our-portfolio-across-the-world (accessed 24 October 2016).

Nordic Environment Finance Corporation (n.d.), "Geographical areas: Europe", webpage, www.nopef.com/cases/europe/ (accessed 24 October 2016).

OECD (2016a), "Project-level Data for every Climate-related Development Finance Project in 2013-14", *Climate Change: OECD DAC External Development Finance Statistics* (database), www.oecd.org/dac/stats/climate-change.htm (accessed 24 October February 2016).

OECD (2016b), "Annex 18. Rio markers", *Revised Definition and Guidance for the Climate Rio Markers*, DCD/DAC(2016)3/ADD2/FINAL, OECD Publishing, Paris, www.oecd.org/dac/environment-development/Annex%2018.%20Rio%20markers.pdf.

OECD (2016c), *Environmental Lending in EU Eastern Partnership Countries, Green Finance and Investment*, OECD Publishing, Paris, http://dx.doi.org/10.1787/9789264252189-en.

OECD (2015a), *Climate Fund Inventory Database* (database), http://qdd.oecd.org/subject.aspx?subject=climatefundinventory (accessed 24 October 2016).

OECD (2015b), "Climate-related development finance in 2013-14" (flyer), OECD Development Assistance Committee, Paris, www.oecd.org/dac/environment-development/Climate-related-dev-finance-ENG.pdf.

Partnership for Market Readiness (2016), "Participants", webpage, www.eurasia.undp.org/content/rbec/en/home/library/environment_energy/renewable-energy-snapshots.html (accessed 24 October 2016).

UNDP (2014), "UNDP in Central Europe and Asia Renewable Energy Snapshots", webpage, www.eurasia.undp.org/content/rbec/en/home/library/environment_energy/renewable-energy-snapshots.html (accessed 24 October 2016).

UNECE and REN21 (2015), "The UNECE Renewable Energy Status Report", United Nations Economic Commission for Europe, Copenhagen and Renewable Energy Policy Network for the 21st Century, Paris, www.unece.org/energywelcome/areas-of-work/renewable-energy/unece-renewable-energy-status-report.html.

UNFCCC (2015), *Adoption of the Paris Agreement*, Decision 1/CP.21, http://unfccc.int/resource/docs/2015/cop21/eng/l09r01.pdf.

World Bank (2016), *World Development Indicators* (database), http://data.worldbank.org/data-catalog/world-development-indicators (accessed 24 October 2016).

World Bank (2014), District Heating Energy Efficiency (P132741), Project Appraisal Document, the World Bank Group, http://www-wds.worldbank.org/external/default/WDSContentServer/WDSP/IB/2014/05/09/000333037_20140509114356/Rendered/PDF/PAD7970REVISED0PUBLIC00R20140009301.pdf (accessed 24 October 2016).

Chapter 3

Assessing readiness to access climate finance

Enhancing "readiness" of the countries of Eastern Europe, the Caucasus and Central Asia (EECCA) to access scaled-up international climate finance is essential to accelerate their transition towards low-carbon and climate-resilient economies. This chapter explores ways in which the EECCA countries and their partners can first assess the countries' readiness to access climate finance, and then identify possible areas for improvement. Based on the review of existing climate finance readiness programmes implemented by several development co-operation partners, this chapter outlines key questions to assess the EECCA countries' climate finance readiness. These questions fall into the following four categories: (a) *planning targets, strategies and policies;* (b) *building institutional capacities;* (c) *developing programmes and projects; and* (d) *implementing, monitoring, evaluating and learning.*

Background

While the countries of Eastern Europe, the Caucasus and Central Asia (EECCA) still need further finance from various sources to achieve their climate targets and enhance their ambition levels over time, the climate finance architecture is becoming increasingly complex. Thus, enhancing "readiness" of the EECCA countries to access scaled-up international climate finance will help them accelerate their transition towards low-carbon and climate-resilient economies. The term "readiness" for climate finance is used here to mean various elements needed to ensure the access of countries to certain climate finance sources, which include relevant policy frameworks, strong institutional and human capacities, and functional monitoring and evaluation systems.

Since improving climate finance readiness can include diverse activities, countries can benefit from first assessing their states of readiness, exploring areas for improvement and identifying their priority actions. In this regard, this chapter explores how to meaningfully assess the readiness of the EECCA countries to access climate finance. The report builds on the programmes or approaches implemented by development co-operation providers (Table 3.1). Based on the review, this chapter outlines key questions for EECCA countries to assess their readiness and to identify priority areas for improvement. In addition, while this report does not solely focus on access to the GCF funding, Box 3.1 outlines key institutions and capacities needed to access its resources as one example.

Table 3.1 shows that several multilateral and bilateral development co-operation partners have already engaged in activities to enhance climate finance readiness. Such institutions include the Green Climate Fund (GCF), the Global Environment Facility (GEF) and other bilateral agencies and multilateral institutions such as Deutsche Gesellschaft für Internationale Zusammenarbeit (GIZ), the KfW Development Bank, the United Nations Environment Programme (UNEP), the United Nations Development Programme (UNDP) and the World Resources Institute (WRI). There are ongoing or planned readiness programmes in Georgia, Kazakhstan, Moldova, Tajikistan and Uzbekistan, some of which are detailed in the next sub-section.

Table 3.1. **Examples of readiness support programmes and approaches**

Entity	Name of programme or approach
Green Climate Fund (GCF)	Readiness and Preparatory Support Programme (the Readiness Programme)
GIZ and the KfW Development Bank	The Climate Finance Readiness Program (CF Ready) (supported by German Ministry for Economic Cooperation and Development)
UNEP, UNDP and WRI	The GCF Readiness Programme (supported by German Ministry for the Environment, Nature Conservation, Building and Nuclear Safety, BMUB)
Global Environment Facility (GEF)	GEF Country Support Programme (CSP)

Sources: GEF(n.d.); GIZ(2013a); OECD(2015a); UNEP, UNDP and WRI(2015); GCF (2016b, 2015).

Box 3.1. Key Actors and their roles in access to the GCF funding resources

The Green Climate Fund (GCF) has been an operating entity of the Financial Mechanism of the UNFCCC since 2014. The GCF is designed to disburse international climate finance from various sources mainly by developed countries (but also emerging and developing countries) to mitigation and adaptation projects in developing countries. There are several types of institutions involved in accessing and using funding from the GCF as outlined below.

National Designated Authorities (NDAs): Each country hosting activities funded by the GCF must nominate an NDA, which co-ordinates the country's engagement with the GCF. The scope of an NDA's work is diverse: convening national stakeholders; strategic oversight of GCF-funded activities to ensure alignment with national priorities; issuance of no-objection letters for project/programme proposals; approval of readiness programme proposals; and nomination of possible national or regional accredited entities for direct access. Regional, national or sub-national institutions that want to be accredited to the GCF need to receive a nomination letter from the NDA in their country or a country in which they intend to operate before applying for accreditation.

Accredited Entities: An accredited entity is a national, regional or international institution responsible for implementing projects supported by the GCF. Individual accredited entities must be accredited by the GCF Board, and for this purpose must show they are capable of, and have experiences with, the following functions:

- identifying, preparing and appraising projects
- overseeing and controlling implementation of an approved project or activity, including monitoring performance, assessing project expenditure against project budget, and reporting on progress made
- monitoring and evaluating implementation of funded activities
- demonstrating risk management capabilities.

National and Regional Accredited Entities: These are the forms of accredited entities to pursue direct access. The GCF promotes the direct access modality by which recipient countries directly access financial resources of the Fund and manage projects. A national or regional accredited entity is nominated by the country's NDA and accredited by the GCF Board.

International Accredited Entities: Countries can also access the GCF resources through international accredited entities, such as UN agencies, multilateral development banks and international financial institutions. International Accredited Entities must also be accredited by the GCF Board, but do not need nomination from NDAs.

As of June 2016, the GCF Board has accredited 33 entities, of which 19 institutions (58%) are international, 5 (15%) are regional and 9 (27%) are national (GCF, 2016a). Entities are accredited based on their experience in implementing a certain category of projects/ programmes. The different categories are listed below:

- Functions: project management, grant award, on-lending, and/or blending of funds
- Project sizes: micro scale (under USD 10 million), small scale (USD 10-50 million), medium scale (USD 50-250 million) and large scale (over USD 250 million)
- Risk categories: low, medium and high risks (based on guidance of the International Financial Corporation).

Box 3.1. **Key Actors and their roles in access to the GCF funding resources** *(continued)*

All accredited entities (international, regional and national) must meet basic fiduciary standards as outlined below, while those wishing to manage certain types of projects must meet additional specialised requirements. In addition, accredited entities must meet environmental and social safeguards, and have experience working on gender and climate change. A summary of criteria appears below. Further details about requirements on those standards, safeguards and track records can be found at the GCF (n.d.).

- clearly defined management and administrative capacities
- a credible financial management and accounting system and a track record of financial statements
- procedures in place for internal and external audits
- internal financial controls to ensure that financial risks are properly managed
- formal procurement standards, guidelines and systems.

Source : GCF (2016a, 2015).

Key elements for climate finance readiness

These existing readiness programmes and approaches often split the process of enhancing a country's climate finance readiness into multiple phases. Figure 3.1 illustrates the possible elements of the process of assessing climate finance readiness. The elements relevant to exploring readiness for climate finance are clustered into: *(a)* planning targets, strategies and policies; *(b)* building institutional capacities; *(c)* developing programmes and projects; and *(d)* implementing, monitoring, evaluating and learning.

Figure 3.1. **Elements of exploring readiness for climate finance**

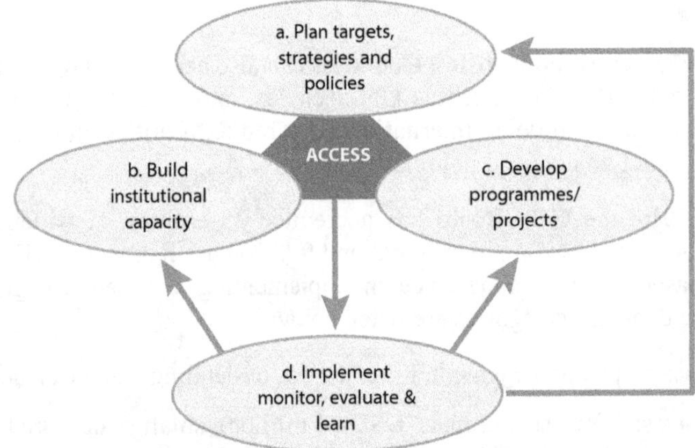

Sources: Adapted from GEF (n.d.); GCF (2016b, 2015); GIZ (2013a); OECD (2015a); UNEP, UNDP and WRI (2015).

Many of the approaches (e.g. GIZ, UNEP/UNDP/WRI and the OECD) set "planning" as a first step of such processes. Planning includes developing climate-related policies and strategies, and integrating them into the broader set of national development policies. To attract not only international public sources, but also domestic finance and private investments, climate and development policies must align with investment, competition and financial policies. Such planning also often includes prioritising of necessary climate actions by the country, taking into account the interests of a range of stakeholders.

Establishing or improving responsible institutions and enhancing their capacities are the core of those existing readiness programmes. Most of the EECCA countries have national designated authorities (or focal points) for the Green Climate Fund and national focal points for the Global Environment Facility. Most of those institutions are governmental bodies in charge of the environment (e.g. a ministry of environment and a state agency for environmental protection). Other ministries (e.g. finance) and public or private entities in the region have also worked with a range of providers of international climate-related development finance.

Potential programmes and projects could be identified and developed in parallel with capacity development for relevant institutions. Funding proposals must clearly articulate how the project or programme is aligned with countries' needs and priorities for national development and climate change strategies (OECD, 2014). Actors that engage in identifying, developing and appraising project or programme proposals vary between countries and depend on how they access climate finance resources. These actors can include national and regional public institutions, private sector entities, and bilateral and multilateral development co-operation partners. Experience shows that feasibility and pre-feasibility studies are often costly, especially in countries where certain historical data on climate risks and records on past projects are not readily available (TWN, 2016).

With such policies, strategies, national priorities and institutions in place, the country may apply for climate finance to fund the projects or programmes in its portfolios, which fit its needs and priorities. International climate-related development finance can be accessed bilaterally or through multilateral agencies. For some multilateral funds, such as the GCF and the Adaptation Fund, countries could directly access the funding resources through national or regional implementing entities that must be accredited by the Board of each fund. Such implementing entities are expected to have the capacity to ensure fiduciary, environmental and social standards in the country, according to the sizes of targeted projects. The GCF, for instance, can provide support to raise the capacity of potential national and regional accredited entities.

Countries can also access such funding through international implementing entities that include multilateral development banks, international financial institutions and the UN agencies. This international access has been a predominant modality to obtain climate finance in many developing countries, including the EECCA; it is often challenging for countries with limited resources to have a domestic institution meet requirements for direct access. After successfully accessing and obtaining climate finance, an international, regional or national implementing entity needs to implement a project or programme in an effective, efficient and accountable manner in the country.

The next phase should be monitoring and evaluation of the progress and effectiveness of the project or programme. Lessons learned from current activities could inform the process of developing new activities and scaling-up and diversifying future finance inflows. Information obtained through monitoring and evaluation will also help meet the reporting requirements of some types of international finance. There is increasing pressure from the public in donor countries to demonstrate the effectiveness of funded activities. Therefore, enhanced capabilities of relevant entities within the countries that seek climate-related development finance in meeting the reporting responsibilities would help to strengthen trust between providers and recipients of finance, and increase possibility to access climate finance (OECD DAC, 2014).

Examples of readiness programmes in Tajikistan and Georgia

The CF Ready Programme is funded by the German Federal Ministry for Economic Cooperation and Development (BMZ) and implemented by GIZ and KfW. GIZ's activities are furthermore co-financed by USAID and the Ministry for Environment of the Czech Republic. The CF Ready Programme supports selected partner countries in accessing financial resources of the GCF and other climate funds and using the funds effectively (GIZ, 2013a).

The GCF Readiness Programme, funded by the German Federal Ministry for the Environment, Nature Conservation, Building and Nuclear Safety (BMUB), is implemented by UNEP, UNDP and WRI. The programme aims to help countries effectively and efficiently plan for, access, manage, deploy and monitor climate financing, including through the Green Climate Fund (UNEP, 2014).

Specific work can differ between countries, reflecting their different priorities and needs. The programme has supported participating countries in establishing national designated authorities (NDAs) to the GCF and strengthening their institutional capacities. An NDA is meant to be in charge of communication and liaison with the GCF, and is the principal signing authority on behalf of the government on matters regarding the Fund. Establishing functional teams in NDAs and strengthening their capacities are primary aims of those readiness programmes. For instance, an NDA is supposed to have capacities to establish and implement a no-objection procedure to ensure consistency of funding proposals with national plans and strategies. NDAs also need to co-ordinate relevant institutions in the country and guide climate finance flows. The programmes also help select and set up national institutions to gain the GCF accreditation in order to access the funds. The aim of such support is to strengthen the institutions so they can support and carry out effective climate programmes in line with national development goals. Some examples from Tajikistan and Georgia on specific work items under the internationally supported readiness programmes follow.

Tajikistan

The GIZ/KfW programme in Tajikistan, which started in 2014, has already produced tangible outcomes. The tasks under the CF Ready Programme for Tajikistan include the following (GIZ, 2015a):

1. Supporting development of a functional NDA: GIZ helped form an NDA team within the Committee on Environmental Protection (CEP). The NDA functions of the CEP have benefited from the experiences and structures that the committee gained through the Pilot Program for Climate Resilience (PPCR) activities under the Climate Investment Funds (see more in the country report for Tajikistan). The GIZ's programme for Tajikistan has envisaged enhancing the following capacities of the NDA (GIZ, 2015a):
    - adequate knowledge of climate-related national priorities, strategies, and plans
    - ability to contribute to and drive national development strategies and plans

- familiarity with both mitigation and adaptation efforts and needs in the country
- familiarity with relevant institutions and stakeholders in the countries (e.g. multilateral and bilateral institutions, civil society organisations, and sub-national, national or regional entities)
- capacity to facilitate country co-ordination mechanisms (detailed in point 3 below) and multi-stakeholder engagement for country consultations
- ability to monitor and evaluate in accordance with relevant guidelines of the Fund.

2. Setting up institutional arrangement of a no-objection procedure: The CEP and GIZ work together to set up a process to issue letters of the no-objection. A no-objection letter should be provided to the GCF Secretariat in conjunction with submission of any funding proposal seeking the Fund's resources. This ensures the proposal is consistent with national strategies, has no conflict with better programmes or projects, and avoids undue harm or costs on any communities or the environment in the country.
3. Setting up a co-ordination mechanism: GIZ and the CEP are also working to establish a process to co-ordinate different ministries, governmental agencies and stakeholders relevant to specific activities to seek the GCF funding. Co-ordination is needed to take into account opinions and interests of all the relevant bodies, and feed into the no-objection process.
4. Improving awareness of GCF procedures: GIZ provided the CEP and other stakeholders in the country with training on climate finance and NDA tasks through the Climate Finance Readiness Training.
5. Supporting development of project pipeline: The KfW Development Bank also supports Tajikistan under the CF Ready Programme, particularly in project development. The work includes conducting pre-feasibility and feasibility studies for specific adaptation or mitigation proposals for the GCF funding.

Georgia

GIZ is also supporting Georgia in enhancing readiness to access climate finance, especially from the GCF. A significant part of the Georgian commitment in its INDC is conditional on international support. Thus, increasing readiness for climate finance, including for the GCF, has become a priority in the implementation phase of its INDC. Building on its support for development of the Georgian INDC, GIZ is developing a draft GCF country programme for Georgia.

Such support includes outlining how priorities and project proposals suitable for the GCF funding can be developed on the basis of the targets and activities expressed in the following: the INDC, existing and planned policy documents such as the Low-Emission Development Strategy, various Nationally Appropriate Mitigation Actions and information included in the Third National Communication to the UNFCCC. Experiences gained through the Covenant of Mayors initiative can also inform the development of project proposals. Such proposals should also be diverse, including mitigation, adaptation and cross-cutting activities (Kindermann, 2015).

Key questions for understanding climate finance readiness

This sub-section outlines key questions for the EECCA countries when they wish to assess their readiness to access further climate finance and identify priority areas for improvement. The questions draw on the existing readiness programmes and approaches in the previous section, and are clustered into four elements: *(a)* planning; *(b)* building capacity; *(c)* developing projects; and *(d)* implementing, monitoring and evaluating progress. Some questions under (b) only relate to those that wish to pursue direct access modalities as indicated.

The lists of questions outlined below are not exhaustive or aimed to illustrate conditions of accessing specific sources of climate finance. Rather, they intend to help the EECCA countries and their development co-operation partners assess their readiness levels. Each of the following four sub-sections lists key questions followed by more detailed discussion.

a. Plan targets, strategies and policies

Key questions to the EECCA governments

- Are the country's needs and priorities for specific mitigation and adaptation actions identified and documented through a broad engagement with relevant stakeholders (e.g. ministries, state agencies, sub-national governments, private sectors, local residents and civil society organisations)?
- Are relevant national-level climate targets, strategies and policies in place to pursue low-carbon and climate-resilient development?
- Are specific actions to implement those strategies and policies already planned and taken?
- Are non-climate policy frameworks such as fiscal, investment promotion, competition and public governance policies in place and conducive to promoting climate actions?
- Is a national- or sector-level financing plan for priority climate actions available?
- Is adequate information on finance sources and access modalities provided, shared and understood among relevant stakeholders within the country in order to implement such a financing plan?
- Is further need for information, capacity, and/or technologies identified in order to implement priority measures? Has a relevant government body (or bodies) started to deal with the need?
- Does the country have predictable and functional enabling environments for attracting private-sector investments such as: policy instruments (e.g. regulatory, economic and information instruments to tackle barriers to investment); necessary capacities of stakeholders (e.g. local banks) to promote investments; and tools for economic assessments and cost-benefit analysis?

The understanding of own needs and circumstances is the first step to prepare for accessing financial resources and allocating them to the most needed climate actions in a country. For this purpose, it is useful to take stock of current climate adaptation and mitigation targets, legal and policy frameworks, and programmes and projects, and where possible, catalogue key lessons learned through related reports.

Regarding the targets, all EECCA countries except Uzbekistan have internationally communicated their GHG mitigation contributions through their Intended Nationally Determined Contributions (INDCs) in time for COP21 in 2015 as outlined in Chapter 1. Adaptation targets are also included in eight INDCs, although the content differs significantly between the countries. Some INDCs (e.g. by Armenia, Azerbaijan, Kyrgyzstan, Tajikistan and Turkmenistan) have explicitly listed the countries' priority climate actions, whereas others (e.g. by Belarus, Georgia and Kazakhstan) refer to other national policy documents or their National Communications to the UNFCCC, which contain the priorities.

It is not clear to what extent those INDCs can be a good basis for actions needed to achieve mitigation and/or adaptation targets. Therefore, further elaboration on the "pathways" towards achievements of the targets is crucial in Nationally Determined Contributions (NDCs) or other relevant policy documents such as Nationally Appropriate Mitigation Actions (NAMAs), National Adaptation Plans (NAPs) and Low-Emission Development Strategies (LEDS), as well as country-specific national climate strategies.

Apart from INDCs, most of the EECCA countries, to a different extent, have developed and implemented a range of climate-related policies (summarised in Table 3.2). Nine countries have quantitative targets on renewable energy or energy saving, or both (UNECE and REN21, 2015). Despite progress over the past decade, common barriers remain to effective policy implementation, including the absence of sufficient tariff pricing; the need for secondary legislation to elaborate on legal, regulatory and financial mechanisms; and lack of clear and enforceable technical rules for grid integration (UNDP, 2014a; IEA, 2015).

Table 3.3 summarises specific examples of domestic policy instruments that can help mobilise climate finance. Improving efficiency in energy generation, transmission and consumption has been a priority in many EECCA countries and significantly financed by international and domestic sources.

Eight EECCA countries have renewable energy targets or economic incentive schemes (e.g. feed-in tariff schemes) to promote the development of renewable energy sources, which in itself represents considerable progress over the last decade. However, many countries have not followed through with significant action. Those countries with renewable targets often lack secondary legislations or bylaws to provide potential investors or creditors with sufficient clarity of the legal, regulatory and financial mechanisms to support project development (UNDP, 2014a; IEA, 2015). The tariffs for renewable energies are often not sufficient to provide economic incentives to potential investors. Belarus, which has used the highest percentage (5.3%) of renewable energy (e.g. biofuels and waste) in its total primary energy supply, is a relatively successful example of promoting renewable energy. Domestic policy incentives such as the relatively high prices put on renewable energy under the feed-in tariff scheme have promoted the introduction of renewable energy in Belarus.

Many EECCA countries have made notable progress in, for instance, reduction of energy intensity per GDP (see Figure 3.2). Nevertheless, the comparison with OECD member countries suggests there is still significant room for improvement and further investments are still needed.

Table 3.2. **Summary of national-level mitigation and adaptation policies in the EECCA region**

	ARM	AZE	BLR	GEO	KAZ	KYR	MDA	TJK	TKM	UKR	UZB
Regulatory policies for renewable energies											
Renewable energy targets	X	X	X		X	X	X	X		X	
Biofuels obligation / mandate			X							X	
Electric utility quotas obligation / Renewable Portfolio Standard (RPS)			X								
Feed-in tariff / premium payments	X	X	X	X	X	X				X	
Heat obligation / mandate											
Net metering	X		X							X	
Tendering (i.e. public bidding) for renewable energy											
Tradable renewable energies certificates			X		X			X			
Fiscal incentives for renewable energies and public financing											
Capital subsidy / rebate					X						
Energy production payment	X	X	X		X	X		X		X	
Investment or production tax credits					X	X		X		X	X
Public investment, loans or grants	X	X	X	X	X	X	X	X			X
Reduction in sales, energy, CO_2, VAT or other taxes	X		X		X	X	X			X	
Energy efficiency policies											
Energy efficiency target			X		X		X	X		X	X
National energy efficiency awareness campaigns		X	X	X	X	X	X	X			X
National energy efficiency regulations, standards or laws		X	X		X		X	X		X	X
Governmental institution(s) to formulate and implement energy efficiency strategies and policies		X	X	X	X		X	X		X	X
Energy efficiency labelling policies			X				X			X	X
Adaptation											
National-level comprehensive policy document that facilitates adaptation						X	X				

Sources: UNFCC (n.d.); adapted from REN21 and UNECE (2015).

Table 3.3. **Examples for policy instruments to mobilise climate finance**

Measure	Country	Descriptions of examples
Public investment in capital formation by state budgets	Belarus	Government support (by national and local government budgets) of USD 1.9 bln for energy efficiency measures in 2011-15, covering 22% of total estimated costs (USD 8.66 bln).
	Uzbekistan	Co-financing from the state, including for energy efficiency and low-carbon designs for rural houses, supported by the GEF (about USD 17 mln from the state budget out of the total cost of USD 114 mln).
	Turkmenistan	The state budget is mainly allocated to the Ministry of Nature Protection and the State Committee for Fisheries.
Direct subsidies for energy efficiency measures	Kazakhstan	50% subsidies for energy audit and implementation of energy management systems based on "Energy Saving 2020" Programme (together with other types of support).
Tax/fiscal incentives	Ukraine	Tax reductions/exemptions for renewable energies, for instance: exemption from import VAT and customs duties for renewable energy equipment 75% reduction in land tax for land used for renewable energy power plants exemption from corporate tax on profit derived from the sale of electricity produced from renewable sources.
	Kyrgyzstan	Imported equipment for the use of renewable electricity is exempt from customs duties (taxes or tariffs) by the Law of the Kyrgyz Republic on Renewable Energy Sources (adopted in 2009).
	Tajikistan	Exemptions from customs duties and VAT for hydropower on imported materials and equipment along with exemption from water royalty tax, profit tax, land tax, capital facility tax and social tax for employees during the construction process.
Feed-in tariff	Ukraine	The Green Tariff, a feed-in tariff scheme for electricity generated from renewables, is open until 1 January 2030 (introduced in 2009 and amended in 2012 and 2015).
	Belarus	The feed-in tariff scheme is applied to solar (EUR 310/MWh) and biomass, biogas, geothermal and wind power (EUR 150/MWh) and hydropower (EUR 130/MWh).
	Azerbaijan	The feed-in tariff scheme is applied to wind (EUR 45/MWh) and small hydro power plants (EUR 25/MWh).
	Armenia	Tariffs for small hydro, wind and biomass power plants.
Power purchase obligation	Georgia	Hydropower plants of up to 100 MW are offered long-term power purchase agreements with Georgia's Energy System Commercial Operator under the Renewable Energy State Program.
Environmental information dissemination	Moldova	Labelling scheme for appliances and equipment about energy consumption performance.

Source: OECD (forthcoming); Liu, H., D. Masera and L. Esser (2013); UNDP (2014a); UNECE and REN21 (2015); adapted from IEA and IRENA (2016).

Figure 3.2. **Total primary energy supply per unit of GDP (tonne of oil-equivalent per thousand 2005 USD)**

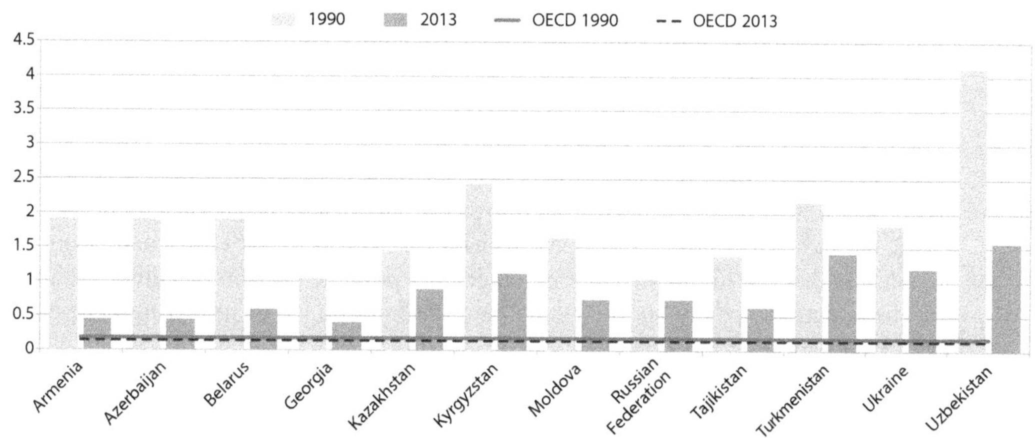

Source: IEA (2016).

In most EECCA countries, policies for adaptation have been less well developed than those for mitigation, which may make it more challenging for them to access finance for adaptation rather than mitigation. While all countries have policies that relate to climate change adaptation (e.g. policies for agricultural, water and/or forestry sectors), only Kyrgyzstan and Moldova have national-level adaptation policy frameworks to date. For instance, the Climate Change Adaptation Strategy of Moldova aims to establish a strong enabling environment and clear direction for an effective and coherent climate change adaptation process across all relevant sectors, and outlines expected implementation costs of each work item. Some others (e.g. Armenia, Belarus, Georgia, Turkmenistan and Tajikistan) have started developing national adaptation plans or strategies. An "ecosystem approach", which Armenia is developing, can be a notable example that outlines measures to adapt to a changing climate, and also seeks to maximise the synergies between mitigation and adaptation actions. There are a range of activities at the international level to support developing countries with establishing and implementing National Adaptation Plans (e.g. the NAP Global Network, and a NAP support window under the GCF), which some of the EECCA countries could benefit from.

In addition to thematic or sector-specific policies, countries need to align different existing and planned policy frameworks with long-term climate targets, while avoiding undue administrative transaction costs and negative side effects of one policy on another (e.g. an energy market reform that leads to increased use of carbon-intensive fossil fuel power plants). For instance, shifting investments from technologies based on fossil fuels to renewable electricity technologies requires a coherent package of policies. These include explicit carbon pricing (e.g. carbon taxes or emissions trading); targeted economic incentives (e.g. feed-in tariffs and public tenders); reform of inefficient fossil-fuel subsidies; and targeted support to innovation through, for instance, public support for research and development (OECD, 2015c, 2016a).

While some of the EECCA countries have developed overarching policy frameworks (examples are listed below), further research could examine the effectiveness of different types of policy measures and the level of alignment among them. Such work could aim to improve access to climate-related development financing sources and/or promote investments in low-carbon technologies, as well as improving the broader investment environments and business conditions in the EECCA countries.

Examples of overarching policy frameworks

- Strategic Development Programme of the Republic of Armenia for 2012-25: This programme, approved in 2014, sets out the consolidated priorities for the socio-economic development of the country, its goals, the main barriers and constraints to development, and the key reforms and policy instruments needed for achieving priority development goals. Achieving these priorities will be in line with addressing issues on environmental protection and the sustainable management of natural resources.
- Azerbaijan 2020: A Look Into The Future: This development concept, approved in 2012, outlines Azerbaijan's development challenges and strategic views, and priorities to address them. It highlights, among others, the possible impacts of climate change on the country's society and economy, and the importance of preparing necessary policy measures.
- The Concept of Kazakhstan for Transition to Green Economy: The concept, approved in 2013, identifies regulatory priorities and measures for green growth. It highlights

a range of recommendations that include adjusting existing laws and regulations to achieve a "green economy" including investment facilitation in clean energy.

- National Strategy for Sustainable Development of the Kyrgyz Republic for 2013-17: This is an overarching strategy that includes the following objectives:
  - improving regulation to encourage energy savings
  - increasing involvement of the state, local authorities and non-governmental organisations in energy savings and energy efficiency in buildings
  - promoting the use of energy-efficient technologies in buildings.

Governance structures to oversee climate actions within a country are often complex. For instance, in a number of internationally supported projects in the energy sectors in the EECCA countries examined by this report, finance ministries tend to be the primary borrowers. Conversely, implementing institutions are often diverse within a country (e.g. an environmental ministry, energy ministry, development ministry, a joint-stock company and a national funding entity). Thus, some of the EECCA countries could usefully explore co-ordination of individual entities and projects at the national level in light of the long-term climate targets and broader development agendas to maximise the benefits of climate-related finance interventions. For example, a dedicated public authority for developing energy efficiency policies and overseeing their implementation could help ensure energy efficiency is a strategic priority and provide greater co-ordination (IEA, 2015).

While current INDCs of seven EECCA countries have targets conditional on international support, most of them have not clarified what and how much support they need. Notably, Kyrgyzstan's and Moldova's INDCs have explicitly mentioned financial needs. Belarus has also estimated the costs of energy efficiency actions within the country in its national policy documents, and Georgia has estimated long-term adaptation costs and communicated them in its INDC. Apart from their INDCs, Azerbaijan, Georgia, Kazakhstan and Moldova (phase I) and Armenia, Turkmenistan and Uzbekistan (phase II) have conducted Technology Needs Assessments (TNAs) under the UNFCCC, which also identified the associated cost. However, these EECCA countries have not yet developed concrete investment planning based on results from TNAs.

The EECCA countries should gather, compile and understand information on international finance sources that are already directed and/or potentially available. The landscape of climate-related development finance is changing day-by-day. Improving information availability and increasing awareness of the multitude of funding sources and instruments can help countries access this funding more easily (OECD, 2015a). Examples of some financial instruments are outlined in Table 3.4.

In light of the country's climate and development needs and policies, analysis could identify priority sectors and geographical areas for financing and the most suitable financial sources. EECCA countries could also usefully compare access modalities for those international climate funding sources with their own national systems for ensuring fiduciary, environmental and social standards, as well as the sizes and types of targeted projects. This would enable the countries to identify the most suitable sources to finance their climate actions.

Table 3.4. **Instruments that can be used to finance climate actions**

Type	Financial instruments and tools	Examples	Mitigation or adaptation	Development level of countries	Local financial market maturity*	Considerations
Public or Private philanthropy	Grants	• Grants for upfront costs • Technical Assistance	Both, but may be more critical for adaptation	With limited institutional capacities or resources	Early market phase	• Can contribute to enhancing readiness to unlock investment • Paticularly suitable for projects with low/ no tangible profits or revenue streams (i.a. concessional instruments) • Can cover/reduce up-front or operating costs, and country contexts, which are too risky to invest • Difficult to determine leveraging of private sector finance (for tracking purposes) • Limited public resources might constrain the extent to which climate finance provision is increased
	Public-sector loans	• Concessional • Non-concessional • Green credit lines	Both (but concessional may be more important for adaptation)			
	De-risking interventions	• Guarantees • Political/Resource risk Insurances	Both			
Fixed incomes (private sector)	Loans	• Green credit lines • Direct/Co-investment • lending • Syndicated project loans	Both			• Particularly suitable for projects with large up-front investment requirements • Many debt securities offer fixed returns for a set period of time, thus are attractive for institutional investor • Bonds can be applied to an array of projects across sectors, thus facilitate scaling up (similar to programmatic approach).
	Bonds	• Project bonds • Green bonds • Sub-sovereign bonds				
Mixed	Hybrid	• Subordinated loans/ Bonds • Mezzanine Finance	Both			
Equity private and public	Listed	• Listed infrastructure and utilities stocks	Mainly mitigation	With greater institutional capacities & resources	Established market phase	• Can be suitable for larger mitigation projects with high risk-return profile • Higher investor risk than other instruments • Can be more complicated to track

* Equity finance can also be provided for early stage technologies through venture capital funds and other sources.

Source: Kato, T., J. Ellis and C. Clapp (2014); adapted from OECD (2015b, 2014c).

b. Build institutional capacity

Key questions to the EECCA governments

- Is domestic institutional set-up for climate change and climate finance clearly defined and understood?
- Is a process in place to co-ordinate relevant stakeholders internally (e.g. across ministries/ state agencies) and externally (e.g. with development co-operation partners, sub-national governments, civil society and the private sector)?
- How is such a process managed? (Ideally a multi-departmental team chaired by a senior official or a minister should manage it.)
- Does a national co-ordination body (e.g. a National Designated Authority or focal point for the GCF, the GEF or the Adaptation Fund) have sufficient knowledge of the following issues?
    - national priorities, strategies, and plans for development and climate policies
    - domestic stakeholders (e.g. contacts with multilateral and bilateral institutions, civil society organisations and potential candidates for sub-national, national or regional implementing entities)
    - international (multilateral, and bilateral) sources of finance
    - different types of financial instruments
    - implementation status and effectiveness of projects/programmes supported by international sources.

[To pursue direct or enhanced direct access under the GCF and the Adaptation Fund]

- Has a candidate for regional, national or sub-national implementing entities within the country been identified? Is there a functioning national funding entity or national climate fund (especially for enhanced direct access)?
- Does such a potential implementing entity have the following capacities?
    - clearly defined legal basis, governance, management and administrative responsibilities and capacity to meet these responsibilities
    - clearly defined profiles for technical staff members
    - financial management and accounting systems
    - procedures for internal and external audits
    - internal financial controls to manage financial risks
    - formal procurement standards, guidelines and systems.
- Does the potential implementing entity have the capacities to meet environmental and social safeguards required by specific climate funds? (Such safeguards are related to working conditions; resource efficiency and pollution prevention; community health and safety; land acquisition and/or resettlement; biodiversity; indigenous peoples; gender; and cultural heritage).
- Is evidence on the above-mentioned capacities documented so it can be submitted to the Fund's secretariat for the accreditation process?
- Does the potential implementing entity have track records on project implementation (according to relevant size, risk and type)?

It is generally recognised that institutional capacities in the EECCA and other developing countries need to be enhanced for them to seize opportunities to better access and absorb international climate finance. However, a variety of necessary capacities are relevant to the countries' efforts to access such finance. Therefore, it is important to identify and understand the countries' own capacity gaps and domestic institutions whose needs for support should be prioritised.

This report has identified diverse sets of actors within a country in the EECCA region, which are involved in implementation of projects and programmes supported by climate-related development finance. Such actors include environmental ministries or agencies; finance ministries (as primary borrowers of donor funding); other line ministries and agencies; sub-national governments; and private or state-owned enterprises (see the country reports for further details). National focal points of the EECCA countries for the UNFCCC and the GEF tend to be the environment ministries. This is the same for the National Designated Authorities (NDAs) of the EECCA countries for the Green Climate Fund. Conversely, more and more countries in other regions have assigned finance or economy ministries as NDAs for the Fund, compared with those countries' national focal points for the GEF.

A broad engagement of relevant stakeholders is good in itself, but examining the efficiency and effectiveness of such arrangements could benefit from further analysis. Some of the EECCA countries (e.g. Armenia, Georgia, Kyrgyzstan, Moldova and Tajikistan) have an entity that co-ordinates climate-related activities within the country. Yet efforts to strengthen capacities and resources of such entities for accessing further climate finance are still at an early stage. Indeed, Armenia's INDC mentions that strengthening the operation of its Intergovernmental Council on Climate Change is a main capacity development need, but specific needs are not available in the document (Government of Armenia, 2015). Creating a new governmental body or restructuring existing institutional arrangements is considered to be "effective", enabling the countries to review multilateral and bilateral sources of finance and, if necessary, consider the accreditation process for certain climate funds. Such an overarching body (or bodies) could also play an important role in enhancing effectiveness in implementation of individual projects in light of the countries' broader climate and development agenda (UNEP, UNDP and WRI, 2015).

For many countries, such efforts to develop institutional capacities would not start from scratch: Tajikistan, for instance, has been engaged in the Pilot Program for Climate Resilience (PPCR) since 2010. The country established an Inter-ministerial Committee (chaired by the Deputy Prime Minister) with representatives of the national committees and sector ministries. The Inter-ministerial Committee has been playing a crucial role in co-ordinating various actors involved in the sub-projects implemented under the PPCR; it has also become an important player for developing project proposals for the GCF funding as well. Uzbekistan's Ministry of Finance has collected information on development co-operation provided by donors and international entities, in collaboration with other line ministries such as the State Committee for Nature Protection.

Direct access modalities are gaining more and more traction as a way to access climate finance with the aim to enhance efficiency and country ownership. While a number of the EECCA countries have accessed relatively new funding mechanisms for climate change actions (see Table 2.1. in Chapter 2), none has done so through direct access modalities for the Adaptation Fund under the Kyoto Protocol, the GEF (only for National Portfolio Formulation Exercise and Convention Reports) or the Green Climate Fund to date (Table 3.5).

Table 3.5. **Access modalities used to obtain funding resources of the Adaptation Fund and the GCF**

		Global	EECCA
Adaptation Fund (2011-15)	International Access	30	3
	Direct Access	11	0
Green Climate Fund (2015-16)	International Access	13	2
	Direct Access	3	0

Source: AF (2015); GCF (2016a).

Access to such funding through international organisations or multilateral institutions continues to have a crucial role in delivering climate-related finance to the EECCA countries. However, direct access could be one possible measure to help them improve efficiency and strengthen ownership over accessing and absorbing financial resources. As one of their structural features, multilateral climate funds are typically trust funds with a limited number of staff members. Many were thus designed to deliver finance through other organisations (e.g. multilateral development banks and UN agencies). For instance, such organisations have been given responsibility for project development, facilitation and management in accessing international public climate finance. This feature often implies a multi-step process to accessing finance, which tends to be time consuming, and suggests that recipient countries may have lower levels of ownership than when directly accessing finance. Direct access is expected to help the countries at least partly tackle such issues.

Nonetheless, direct access requires strong national institutions that can meet robust fiduciary standards and environmental and social safeguards (Kato, Ellis and Clapp, 2014); a process to get the country's entity accredited may become lengthy. In case of the Adaptation Fund and the GCF, the time needed for accreditation ranged from 5 months for Uruguay to 30 months for Namibia and Kenya (Masullo et al., 2015).

For EECCA countries and many developing countries alike, the exact implications of enhancing their institutional capacities to meet those necessary procedures, standards and safeguards are not easily understood. Indeed, such capacities can be diverse. An implementing entity that can access GCF funding directly needs clearly defined management and administrative capacities; a financial management and accounting system; procedures for internal and external audits; internal financial controls to manage financial risks; and formal procurement standards, guidelines and systems, among others (see also Box 3.1).

An implementing entity is meant to have experience with project management and provide its track record on preparing and appraising projects; overseeing and controlling their implementation; monitoring, evaluating and reporting; and managing project-level risks, among others. In the case of the GCF, an implementing entity is accredited in accordance with project sizes, fiduciary functions and level of financial risks with which it has relevant experiences. For instance, an entity can only receive direct funding from the GCF for projects or activities that are the same size (measured by budget) or smaller than ones previously undertaken. Given such preconditions and the lack of experience of the EECCA countries in direct access, it is likely that any accreditation for entities in the region will be for small-scale and relatively low-risk projects.

c. Develop programmes and projects

Key questions to the EECCA governments

- Is there an initial pipeline of projects and/or programmes for funding proposals that is well-developed in light of national development and climate priorities?
- Are (potential) project proponents (e.g. international and domestic financial institutions, implementing entities, private sector project developers, contractors, civil society organisations, etc.) identified and their roles clearly understood by relevant stakeholders who can be involved in climate actions?
- Are project proponents aware of national needs and priorities so they can justify why the proposed project or programme needs to be implemented in light of these priorities and needs when they apply for specific climate funds or other types of financial sources?
- Is there sufficient information within the project proponents on necessary technologies, regulations and policies, financial instruments and relevant entities that can affect the development of the particular project/programme?
- Do the project proponents have sufficient capacity to identify and appraise project/programme proposals (e.g. knowledge base on conducting cost-benefit analysis and project risk assessment, understanding of key result indicators against which the project is evaluated, etc.)?
- Are track records of past similar projects and programmes readily available to those who need them?
- Is a project risk reduction package available to project proponents (e.g. guarantees and insurance products; support for tariffs; policies for favourable power purchasing agreements; public investment funds; and loan loss reserves, possibly with a technical and financial support of development co-operation providers)?
- Are local stakeholders aware of economic opportunities from various low-carbon and/or climate-resilient actions and available financial resources (e.g. domestic financial institutions, enterprises of various sizes, including small and medium-sized enterprises, other domestic project developers, and construction tender and maintenance contractors)?

In 2013 and 2014 alone, bilateral and multilateral financial providers committed to more than 660 projects in the EECCA region; many of these countries also funded a number of climate-related projects from their own state and municipal budgets (e.g. see sections on domestic climate finance mechanisms in the country reports on Azerbaijan, Belarus, Kazakhstan and Uzbekistan, among others). However, there is still a significant level of unexploited potential for energy saving and renewable energy measures in the region (IEA, 2015). There is also a substantial funding gap for adaptation, despite the assessed high vulnerability to climate change in countries such as Kyrgyzstan, Moldova and Tajikistan.

Despite significant progress in many developing countries over the past years, lack of technical capacity to design and develop project or programme proposals often constrains access to climate finance. Experience of the EECCA countries shows that finding and developing bankable projects on climate mitigation or adaptation has been particularly challenging for small and medium-sized enterprises (EBRD, 2014; OECD, 2016b). For instance, limited capacity to present the necessary inputs, planned activities and expected outputs, outcomes and impacts of project/programme ideas in the form of each fund's "logical framework" can make it challenging for countries to access financial resources of such funds. Further, local commercial banks in these countries tend not to have sufficient technical capacity to assess projects and technologies to be employed, especially those not common in the countries (e.g. large-scale renewable power projects) (CIF, 2014). The GIZ's CF Climate Finance Reflection Tool also mentions that at a project/programme level, technical know-how to identify the appropriate technologies, capacities in project planning and development, and financial expertise to examine and improve costs and returns for the project are all necessary to attract public or private investors (GIZ, 2015).

The governments of the EECCA countries can facilitate further development of project pipelines in various ways. Support for enhancing the evidence base on technologies and financial instruments for decision making at a project/programme level can be an option. The governments can also provide risk-reduction measures for project proponents with possible support of international sources; this can help them promote development of investment environments that are stable and conducive to mobilising further finance for climate-related action from private and public sectors. More detailed discussions follow below.

For project proponents to analyse the "bankability" of their proposals, the governments could support data improvements, such as development of key result indicators and collection of baseline data. To support such activities, the government or the other public and private entities can collect and collate track records of past and ongoing projects. The governments can also make the records publicly available to the extent possible for those planning new projects/programmes. The lack of a track record on (pilot) projects and programmes may inhibit countries from developing their project/programme pipelines efficiently, and also from scaling-up the sizes of existing (or pilot) activities.

Governments or their development co-operation partners have also organised a number of dialogues to share experiences on climate-related projects, which could also help catalyse finance for climate actions at the national, sub-national and sectoral levels. Such information-sharing can be useful to identify practices that are "good enough" to potentially attract climate finance sources (e.g. bankable). They could also help raise awareness among local entities (e.g. financial institutions, enterprises of various sizes, project developers, construction tenders and maintenance contractors) of *(i)* economic opportunities from adaptation and mitigation actions; and *(ii)* national development and climate policy priorities. The latter is particularly important when a project proponent applies for certain types of climate funds that require rationales for the proposed projects in light of national needs and priorities. Care should be taken, however, to ensure these information-sharing opportunities do not become just "talking shops" (Naidoo, 2014).

Providing financial risk management instruments can promote the pipeline development mainly for large infrastructure investments by improving risk-adjusted returns on investment of a project or programme. Risk mitigants and transaction enablers[1] can be powerful tools for promoting mitigation- and adaptation-related investments. Risk mitigants are used to reduce or re-assign investment risks, and include guarantees and insurance products, as well as other types of credit enhancement instruments. Transaction enablers are used to reduce the transaction costs associated with investment, and include securitisation and co-investment platforms.

Some EECCA countries have already used some of those financial instruments. For instance, the EBRD extended a range of credit lines to local banks in EECCA countries with sovereign guarantees as the participating local banks had small capital bases and lacked necessary track records and appropriate knowledge of new technologies (see Table 2.1. in Chapter 2 and OECD 2016b for more details). The EBRD has also provided lending with sub-sovereign (i.e. municipality) guarantees to urban infrastructure projects in Ukraine and elsewhere. In another example, a project supported by the GEF and UNDP in Kazakhstan aims to take a sector-wide, rather than project-based, approach to unlocking private sector investments in renewable energy; it is based on qualitative and quantitative assessment of risks from a private sector perspective. One component of the project plans to provide both technical assistance and direct financial support through financial de-risking instruments and direct financial incentives (GEF and UNDP, 2015).

All the EECCA countries have also put in place national funding entities that provide finance, or co-finance with international sources, to climate-related projects and programmes (Box 3.2). Such entities can help access, blend and co-ordinate various climate-related development finance (both public and private), as well as produce reports to the funding sources. For instance, the Georgian Energy Development Fund (GEDF) identifies new renewable energy projects that will be offered to potential investors (with or without GEDF co-investment). GEDF has invested in 10 projects since 2011 worth over USD 20 million (GEDF, n.d.).

Box 3.2. **National funding entities in the EECCA countries**

The experience of many developing countries shows the significant benefits of using in-country systems to manage activities supported by international and domestic climate-related finance. Such benefits could include reducing duplication and transaction costs; enhancing a country's ownership over results of financing; strengthening linkages between climate policies and the country's core planning and budgeting processes; and improving accountability and transparency.

All the EECCA countries have national funding entities that provide finance, or co-finance with international financial resources, to climate-related projects and programmes. In some countries, such as Moldova and Kazakhstan, different ministries manage several national funding entities. Table 3.6 shows a non-exhaustive list of such national funding entities whose financial resources are/will be at least partly used for climate mitigation or adaptation actions in the EECCA countries.

The aims of using such national funding entities are also diverse. They include to meet capital investment needs for climate-related policies and national development plans; to attract foreign direct investment and domestic private-sector financing; and to co-finance projects supported by providers of international and domestic climate finance. Whether and how investments by such national funding entities are efficiently and effectively implemented deserve further analysis.

Some of these national funding entities could be well-positioned to strengthen country ownership and to enhance coherence among various activities supported by international financial sources, as well as their effectiveness. For instance, the Renewable Resources and Energy Efficiency Fund of Armenia (R2E2) provides loans and grants for renewable energy and energy efficiency projects with the support of funds from the World Bank and the GEF in line with strategic priorities set out by the Armenian government. Some of the national funds, such as Uzbekistan's Fund for Reconstruction and Development, have a broader set of objectives to co-finance projects selected as a priority by the government. If co-ordinated properly with relevant stakeholders, these funds can help countries harmonise efforts to avoid inconsistencies and maximise benefits of projects or programmes for climate actions and broader sets of sustainable development. GIZ helps countries establish and operate national climate funds with appropriate objectives and strategies, organisational structure, project cycle and procedures, as well as the monitoring and evaluation of results.

Table 3.6. **Examples of national entities in the EECCA countries**

Country	National Funding entity	Description
Armenia	The Renewable Resources and Energy Efficiency Fund of Armenia (R2E2)	The R2E2 Fund implements loans and grants for projects with support of funds from the World Bank and the GEF to develop renewable energy and energy efficiency in line with strategic priorities set out by the Government of Armenia. Armenia's INDC also mentions plans for a new revolving fund for climate actions.
Azerbaijan	The State Oil Fund of the Republic of Azerbaijan (SOFAZ)	SOFAZ, established in 1999, aims to transform hydrocarbon reserves into financial assets generating sustainable income. SOFAZ has invested in energy sector infrastructure, water supply and sanitation, irrigation systems and transport, among others. It also provides co-financing for projects supported by international donors.
Belarus	The Innovation Fund of the Ministry of Energy	This fund supports energy-saving measures and is managed by the Ministry of Energy, while the other ministries also have sectoral innovation funds (ECS, 2013).
Georgia	The Georgian Energy Development Fund (GEDF)	The GEDF is a joint stock company established in 2010 by the Government of Georgia to facilitate investment and development of the country's renewable energy sector.
Kazakhstan	Samruk-Kazyna	The Sovereign Welfare Fund Samruk-Kazyna provides financing to the energy sector, either by direct investment in electricity generation and supply facilities, or as a shareholder of national development institutions and national companies.
Kyrgyzstan	The National Fund for Environment Protection	This fund delivers state budget support to a broad set of environmental issues and sustainable forest management. For instance, the Fund allocated 17 million soms (about USD 0.224 million) to energy efficiency projects in 2015.
Moldova	The Energy Efficiency Fund	The Energy Efficiency Fund promotes investment in energy-saving measures, as well as renewable energy projects. It operates based on the state budget, supporting a range of domestically and internationally funded projects. The country also has a range of Environmental Protection Funds.
Tajikistan	The State Environmental Protection Fund	This fund was established by the government and operated under the Committee for Environmental Protection. Recent activities financed by this fund are not clear.
Turkmenistan	The Forest Fund	This fund is managed under the Ministry for Nature Protection to provide financing to sustainable forest management and grazing.
Ukraine	The Energy Efficiency Fund (planned)	This fund plans to provide financing for energy efficiency measures primarily in the residential and public sectors, but also in enterprises of the heat supply industry.
Uzbekistan	The Fund for Reconstruction and Development	Between its creation in 2006 and 2014, the Fund accumulated USD 15 billion in assets in a range of projects and entities related to development activities.

Source: GEDF (2014); Samruk-Kazyna (n.d.); UNECE (2012); ECS (2013); R2E2 (2014); UNDP (2014b); Government of Ukraine (2015); World Bank (2015); AKI Press (2016); OECD (2016c); SOFAZ (2016).

d. Implement, monitor, evaluate and learn

Key questions to the EECCA governments

- Is an accountability mechanism to monitor and report the use of funding in place within a government and properly functioning?
- Have key result indicators been developed to monitor and evaluate the effectiveness of climate finance interventions (e.g. actual GHG emission reductions; installed capacity of clean energy; enhanced adaptive capacities such as the uptake of climate-resilient infrastructure, early warning systems and crop insurances; avoided economic and non-economic losses; development of new technologies; and created employment)?
- Is there a process to integrate obtained information into management processes of government agencies?
- Is there a public database to share relevant information on mitigation and adaptation actions and their results within and outside of the country? Is the database updated periodically?
- Are there regular communication channels with international and domestic climate finance sources about national climate policy needs and priorities that may be updated by using lessons learned through iterative monitoring and evaluation?

There is increasing pressure from the public in donor countries to demonstrate effectiveness of funded activities (OECD DAC, 2014), and to ensure transparency in use of financial resources. Therefore, building capabilities of relevant institutions within the EECCA countries to meet the monitoring and reporting responsibilities would help strengthen trust with providers of finance. Enhanced trust is also likely to help increase the possibility for the countries to access further climate finance in future.

Ensuring effective and efficient spending of financial resources involves various elements. They include an accountability mechanism to track allocation and use of resources; proper integrity management systems; appropriate procurement guidelines under the public finance system within the country; and environmental, financial and economic criteria to appraise and review projects. For reporting, the necessary data must be available to report to external bodies (GIZ, 2013a). The Pilot Program for Climate Resilience in Tajikistan (Phase I), for instance, has conducted a stocktaking exercise to identify capacity gaps in monitoring and reporting practice, among others.

In addition, monitoring and evaluation can produce valuable information needed to understand what has and has not worked in funded activities; this can help improve future pipeline development and access to necessary finance. Such information, for example, could target the environmental performance of the project/programme. This would require key indicators and sufficient data availability to monitor implementation and results. Information on technologies used by the project (or programme) is also useful for scaling up. Those who plan similar projects can benefit since scalability and replicability of projects can highly depend on the quality and maturity of the technologies. Better information on technologies can then lead to more informed investment decision making by actors such as government agencies and financial institutions. Some internationally supported projects in the EECCA explicitly aimed such demonstration effects of the first-of-its-kind technologies in the particular countries. This includes, for example, renewable energy projects co-financed by the Clean Technology Fund in Kazakhstan and Ukraine where there were no previous track records on such types of projects. For similar reasons, information on the financial structures of climate finance interventions can be useful: types of financial instruments to be used depend on economic and financial barriers of individual countries to ensure the financial sustainability of the interventions.

Obtained information needs to be integrated into management processes of the EECCA governments, such as for strategic planning, policy formulation, project or programme management, budget management, and human resource management (OECD DAC, 2014). This can be done by, for instance, raising awareness of senior management, and integrating contributions to monitoring and evaluation into routine tasks of staff where possible. A public database could usefully share the above-mentioned different types of information on implementation and results of climate actions inside and outside the country.

Further work in this area could include analysis of different aspects of "effectiveness" of climate-related development finance. A framework, for example, could examine different types of information on results mainly from completed projects. Such information could be focused on climate-related impacts or outcomes of projects, technologies employed, mobilisation of private sector financing and the relation between domestic enabling environments (e.g. climate and investment policies). This type of analysis could provide a solid basis for discussing various purposes such as scaling-up existing/pilot projects, accessing a broader set of financial sources, and improving the climate and investment policy frameworks for green financing.

Note

1. "Risk mitigants" and "transaction enablers" are terms used in OECD (2015b), "Mapping of instruments and incentives for infrastructure funding: A taxonomy", *Report to G20 Finance Ministers and Central Bank Governors.*

References

AF (2015), *Independent Evaluation of the Adaptation Fund: First Phase Evaluation Report*, Adaptation Fund, www.adaptation-fund.org/wp-content/uploads/2015/11/TANGO-ODI-Evaluation-of-the-AF_final-report.pdf.

AKI Press (2016), "Environmental protection fund allocates $224 thousand for replacement of filters of Bishkek boiler stations", 1 January 2016, www.akipress.com/news:571230/ (accessed 5 September 2016).

CIF (2014), "Investment Plan for Armenia, Republic of Armenia Scaling Up Renewable Energy Program (SREP)", www-cif.climateinvestmentfunds.org/sites/default/files/meeting-documents/armenia_srep_investment_plan_final_0.pdf.

EBRD (2014), "CEEP – Demirbank Azerbaijan", webpage, www.ebrd.com/work-with-us/projects/psd/ceep---demirbank-azerbaijan.html (accessed 24 October 2016).

EC (n.d.), International Cooperation and Development, Blending, Fruit Garden of Moldova", webpage, https://ec.europa.eu/europeaid/blending/fruit-garden-moldova_en (accessed 24 October 2016).

ECS (2013), *In-Depth Review of Energy Efficiency Policy of Belarus*, Energy Charter Secretariat, Brussels, Belgium, www.energycharter.org/fileadmin/DocumentsMedia/IDEER/IDEER-Belarus_2013_en.pdf (accessed 25 August 2016).

GCF (2016a), "Accredited Entities" webpage, www.greenclimate.fund/partners/accredited-entities (accessed 24 October 2016).

GCF (2016b), *Progress and Outlook Report of the Readiness and Preparatory Support Programme*, Green Climate Fund, Songdo, Republic of Korea, www.greenclimate.fund/documents/20182/226888/GCF_B.13_24_-_Progress_and_outlook_report_of_the_Readiness_and_Preparatory_Support_Programme.pdf/ba611e9a-d890-4964-8a45-2e4896341059 (accessed 24 October 2016).

GCF (2015), READINESS Fine print website, www.greenclimate.fund/ventures/readiness/fine-print (accessed 24 October 2016).

GEDF (2014), *Georgian Energy Development Fund Company Presentation*, JSC Georgian Energy Development Fund, http://gedf.com.ge/wp-content/uploads/2014/06/CCGT-GEDF.pdf (accessed 24 October 2016).

GEF (n.d.), "Country Support Programme", webpage, https://www.thegef.org/topics/country-support-programme (accessed 24 October 2016).

GEF and UNDP (2015), "De-risking Renewable Energy Investment", webpage, www.thegef.org/project/de-risking-renewable-energy-investment (accessed 24 October 2016).

GIZ (2015a), *Climate Finance Readiness Programme in Tajikistan – CF Ready*, Deutsche Gesellschaft für Internationale Zusammenarbeit, GmbH, Bonn and Eschborn, Germany, http://naturalresources-centralasia.org/assets/files/EbA_Conference/Climate_Group/666.pdf (accessed 24 October 2016).

GIZ (2013a), *Assessing Needs for Climate Finance Readiness*, Deutsche Gesellschaft für Internationale Zusammenarbeit, GmbH, Bonn and Eschborn, Germany, www.giz.de/expertise/downloads/giz2013-en-climate-finance-reflection-tool.pdf (accessed 24 October 2016).

GIZ (2013b), *Economic Approaches for Assessing Climate Change Adaptation Options Under Uncertainty*, Deutsche Gesellschaft für Internationale Zusammenarbeit (GIZ) GmbH, Bonn and Eschborn, https://gc21.giz.de/ibt/var/app/wp342deP/1443/wp-content/uploads/filebase/ms/mainstreaming-guides-manuals-reports/Economic_assessment_of_CC_adaptation_options_-_GIZ_2013.pdf.

Government of Armenia (2015), Intended Nationally Determined Contribution, www4.unfccc.int/submissions/INDC/Published%20Documents/Armenia/1/INDC-Armenia.pdf.

Government of Armenia (2014), State Agency on Alternative and Renewable Energy Sources website, area.gov.az/azerbaijan-goverment/?lang=en (accessed 24 October 2016).

Government of Ukraine (2015), "Ukraine Invites Sweden to Join Establishment of Energy Efficiency Fund", webpage, www.kmu.gov.ua/control/en/publish/article?art_id=248672500&cat_id=244314975 (accessed 24 October 2016).

IEA (2016), "World Indicators", *IEA World Energy Statistics and Balances* (database), http://dx.doi.org/10.1787/data-00514-en (accessed 30 May 2016).

IEA (2015), *Energy Policies Beyond IEA Countries: Caspian and Black Sea Regions 2015*, IEA/OECD Publishing, Paris, http://dx.doi.org/10.1787/9789264228719-en.

IEA and IRENA (2016), *IEA/IRENA Global Renewable Energy Policies and Measures Database* (database), www.iea.org/policiesandmeasures/renewableenergy/ (accessed 24 October 2016).

Kato, T., J. Ellis and C. Clapp (2014), "The role of the 2015 agreement in mobilising climate finance", *OECD/IEA Climate Change Expert Group Papers*, No. 2014/07, OECD Publishing, Paris, http://dx.doi.org/10.1787/5js03h9ztlbr-en.

Kindermann, U. (2015), "Preparing for the implementation of the Georgian INDC, available support from German Development Cooperation", Presentation at Stakeholders Consultation Meeting on Georgia's Intended Nationally Determined Contribution, Tbilisi, 17 August 2015, http://1067656943.n159491.test.prositehosting.co.uk/wp-content-sec/uploads/2015/08/2015-08-17-INDC-implement-support-GIZ-UK.pdf (accessed 24 October 2016).

Liu, H., D. Masera and L. Esser (eds.) (2013), *World Small Hydropower Development Report 2013*, United Nations Industrial Development Organization, Vienna, and International Center on Small Hydro Power, Hangzhou, China, www.smallhydroworld.org/fileadmin/user_upload/pdf/Asia_Western/WSHPDR_2013_Georgia.pdf.

Masullo, I. et al. (2015), "'Direct access' to climate finance: Lessons learned by national institutions", *Working Paper*, November 2015, World Resources Institute, Washington, DC, www.wri.org/publication/direct-access.

Naidoo, C. (2014), "Green climate fund: Overview and readiness opportunities for Africa", presentation to climate finance readiness leadership programme, August 2014, http://e3g.org/docs/Day_1_CFRL_GCF_Readiness_07082014.pdf (accessed 24 October 2016).

OECD (forthcoming), *Inventory of Energy Subsidies in the EU's Eastern Partnership Countries: Belarus.*

OECD (2016a), *OECD Business and Finance Outlook 2016*, OECD Publishing, Paris, http://dx.doi.org/10.1787/9789264257573-en.

OECD (2016b), *Environmental Lending in EU Eastern Partnership Countries, Green Finance and Investment*, OECD Publishing, Paris, http://dx.doi.org/10.1787/9789264252189-en.

OECD (2016c), *Reforming Economic Instruments for Water Resources Management in Kyrgyzstan*, OECD Publishing, Paris, http://dx.doi.org/10.1787/9789264249363-en.

OECD (2015a), "Toolkit to enhance access to adaptation finance: For developing countries that are vulnerable to adverse effects of climate change, including LIDCs, SIDS and African states", OECD Publishing, Paris, www.oecd.org/environment/cc/Toolkit%20to%20Enhance%20Access%20to%20Adaptation%20Finance.pdf (accessed 24 October 2016).

OECD (2015b), "Mapping of instruments and incentives for infrastructure financing: A taxonomy", *Report to G20 Finance Ministers and Central Bank Governors*, OECD Publishing, Paris, www.oecd.org/finance/private-pensions/Mapping-Instruments-Incentives-Infrastructure-Financing-Taxonomy.pdf.

OECD (2015c), *Aligning Policies for a Low-carbon Economy*, OECD Publishing, Paris, http://dx.doi.org/10.1787/9789264233294-en.

OECD (2014), *Climate Resilience in Development Planning: Experiences in Colombia and Ethiopia*, OECD Publishing, Paris, http://dx.doi.org/10.1787/9789264209503-en.

OECD DAC (2014), *Measuring and Managing Results in Development Co-operation: A Review of Challenges and Practices among DAC Members and Observers*, OECD Publishing, Paris, www.oecd.org/dac/peer-reviews/Measuring-and-managing-results.pdf (accessed 4 July 2015).

R2E2 (2014), "Scaling Up Renewable Energy Program (SREP): Investment Plan for Armenia", webpage, www.r2e2.am/wp-content/uploads/2014/08/Armenia-SREP_2014.pdf (accessed 5 September 2016).

Samruk-Kazyna (n.d.), Informational and analytical portal of Samruk-Kazyna website, http://sk.kz/ (accessed 5 September 2016).

SOFAZ (2016), State Oil Fund of the Republic of Azerbaijan website, www.oilfund.az/en_US/hesabatlar-ve-statistika/son-reqemler.asp (accessed 1 June 2016).

TWN (2016), "GCF Board approves simplified process for small-scale funding proposals and other funding related decisions", TWN Information Service on Climate Change, 5 July 2016, Third World Network, www.twn.my/title2/climate/info.service/2016/cc160705.htm (accessed 24 October 2016).

UNEP (2014), "Green Climate Fund Readiness Programme", webpage, www.unep.org/energy/Projects/Project/tabid/131381/language/en-US/Default.aspx?p=072a4233-85db-49aa-8181-b80e930a050d (accessed 24 October 2016).

UNDP (2014a), "UNDP in Central Europe and Asia, Renewable Energy Snapshots", webpage, www.eurasia.undp.org/content/rbec/en/home/library/environment_energy/renewable-energy-snapshots.html (accessed 24 October 2016).

UNDP (2014b), "ESCO Moldova – Transforming the market for Urban Energy Efficiency in Moldova by introducing Energy Service Companies, Project Document", United Nations Development Programme, www.md.undp.org/content/dam/moldova/docs/Project%20Documents/ESCO%20Moldova%20Project%20Document_EN.pdf.

UNEP, UNDP and WRI (2015), Supporting Readiness Programme website, www.gcfreadinessprogramme.org (accessed 24 October 2016).

UNECE and REN21 (2015), "The UNECE Renewable Energy Status Report", United Nations Economic Commission for Europe, Copenhagen and Renewable Energy Policy Network for the 21st Century, Paris, www.unece.org/energywelcome/areas-of-work/renewable-energy/unece-renewable-energy-status-report.html.

UNECE (2012), *Tajikistan Environmental Performance Reviews*, United Nations Economic Commission for Europe, www.unece.org/fileadmin/DAM/env/epr/epr_studies/TajikistanII.pdf.

UNFCCC (2015), INDCs as communicated by Parties, http://www4.unfccc.int/submissions/INDC/Submission%20Pages/submissions.aspx (accessed 24 October 2016).

World Bank (2015), "World Bank Group – Uzbekistan Partnership: Country Program Snapshot", World Bank, Washington, DC, http://pubdocs.worldbank.org/pubdocs/publicdoc/2015/10/420141446025649239/Uzbekistan-Snapshot.pdf (accessed 24 October 2016).

Chapter 4

Country profiles

*This chapter summarises the country-specific reports on climate-related development finance for 11 countries of Eastern Europe, the Caucasus and Central Asia (EECCA). The full country reports are available on the website of the OECD-hosted GREEN Action Programme [*www.oecd.org/environment/outreach/eap-tf.htm*]. Each report analyses the country's climate targets and priority sectors/areas for climate actions; development finance flows to support climate actions in the EECCA region; and in-country enabling environments, such as laws, regulations, institutional arrangements and domestic financing mechanisms.*

Analytical framework

This chapter summarises country-specific reports for the 11 countries of Eastern Europe, the Caucasus and Central Asia (EECCA) on climate-related development finance. The full country reports are available on the website of the OECD-hosted GREEN Action Programme.[1] Each of the full country reports contains:

- An analysis of the country's climate targets and priority sectors/areas for climate actions based on the country's Intended Nationally Determined Contribution (INDC) if relevant, and other relevant policy documents and reports to the United Nations Framework Convention on Climate Change (UNFCCC).
- An overview of international development finance flows to support the country's climate actions, based on a quantitative analysis for the two years between 2013-14 and a qualitative analysis for 2011-15. This analysis does not offer a complete picture of climate finance from all possible sources. However, it provides a clearer understanding of international (public) finance flows in terms of major sectors/areas, providers and financing structures for individual projects, as well as domestic institutions involved in accessing and using such finance, on which relevant data tend to be scattered.
- Brief overview of in-country enabling environments, such as laws, regulations, institutional arrangements and domestic financing mechanisms, which directly or indirectly relate to promoting low-carbon, climate-resilient development. This analysis is based on publicly available documents on legal and policy frameworks, as well as public financing entities.

The quantitative analysis for 2013-14 used the database from the OECD Development Assistance Committee (DAC) Creditor Reporting System (CRS).[2] This database allows for an approximate quantification of climate-related development finance flows that target climate mitigation or adaptation objectives of the Rio Conventions. The bilateral sources include the OECD DAC members, whereas multilateral sources include multilateral development banks (MDBs), international climate funds and some South-South co-operation and non-DAC member contributions. This study tracks development finance flows for activities that target climate mitigation or adaptation as either their principal or significant objective. The qualitative analysis for 2011-15 is based on project-level information (e.g. project appraisal documents, interim or terminal evaluation reports, and periodic donor reports). Such information is obtained from publicly available information on bilateral and multilateral providers of support.

The DAC CRS records face values of the activities on the dates when recipients sign grant or loan agreements (i.e. commitment, but not disbursement of funds). Therefore, there may be gaps between results from the DAC CRS and recipient countries' external climate-related development finance statistics on the ground, especially when observed over a longer period.

Data sources for both the quantitative and qualitative analysis sections are limited to the OECD DAC member countries, the MDBs and climate funds. Therefore, the sources do not include some non-DAC member donors such as the People's Republic of China and the Russian Federation, or the private sector, which are likely to have provided a significant amount of finance to some of the EECCA countries.

Armenia

Armenia submitted its INDC in 2015, highlighting the need to address both adaptation and mitigation. The country communicated its intention to set the total aggregate quantitative contribution equal to 633 million tCO_{2e} for 2015-50 or an annual average of 5.4 tCO_{2e} per capita (Government of Armenia, 2015). The INDC also states the country will pursue an "ecosystem-based approach" to adapting to climate change. Armenia has already developed a range of legal and policy frameworks on addressing issues concerning climate change and a wider sustainable development agenda (e.g. Strategic Development Programme of the Republic of Armenia for 2012-25) (Government of Armenia, 2016).

During the two years of 2013-14, nearly USD 200 million per year of climate-related development finance was committed to support mitigation and adaptation in Armenia. The level of the committed amount was lower than the average of USD 303 million per year among EECCA countries. Nevertheless, annual climate-related development finance "per capita" committed to the country (approximately USD 55) is about double the EECCA average of USD 27 per capita annually. The allocation of funds between mitigation, adaptation and multi-focal (i.e. both mitigation and adaptation) projects was relatively well balanced (29%, 38% and 33% respectively) during the period. The energy-, agriculture- and water-related sectors got the largest committed amounts of climate-related development finance in 2013 and 2014 (approximately USD 162 million per year, or 67.8%) in the country.

Both bilateral and multilateral providers committed significant amounts of climate-related development finance in 2013 and 2014. The largest contributor was Germany, as well as the World Bank Group (WBG), the Asian Development Bank (ADB) and the European Bank for Reconstruction and Development (ABRD). Loans are predominantly used as financial instruments. Diverse financial instruments are used, including grants, concessional and non-concessional loans and equity.

The Ministry of Nature Protection is involved in a range of climate-related projects supported by international sources. However, many other ministries and governmental agencies, as well as domestic public financing mechanisms, also engage in or co-finance such projects. For instance, the Ministry of Energy and Natural Resources is responsible for energy policy. The Energy Saving and Renewable Energy Fund (R2E2) co-finances projects supported by international sources (R2E2, 2014).

Figure 4.1. **Climate-related development finance flows committed in 2013-14 (Armenia)**

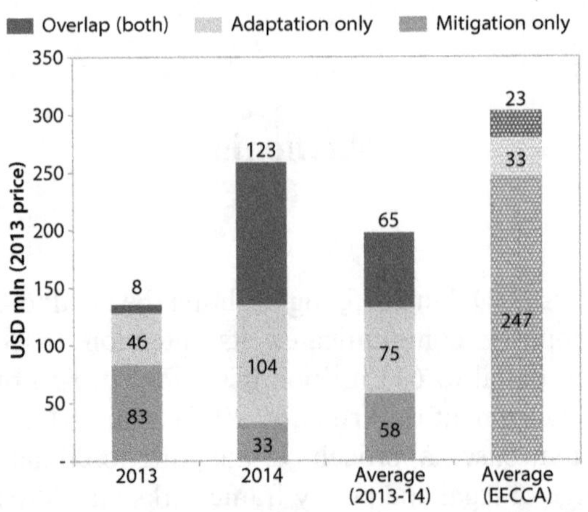

Figure 4.2. **Top 5 Sectors in 2013-14**

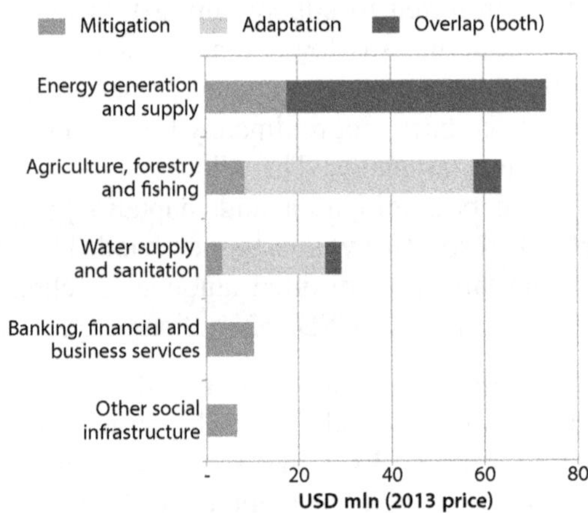

Note: Total climate-related development finance = Mitigation + Adaptation – Overlap (both).

Source: Based on OECD (2016).

Azerbaijan

Azerbaijan submitted its INDC in 2015 with the quantitative targets to reduce total greenhouse gas (GHG) emissions by 25.7 million tCO_{2e} (excluding land use, land-use change and forestry, or LULUCF) or 24.2 million tCO_2e (including LULUCF) by 2030 compared to the 1990 level (Government of Azerbaijan, 2015). The INDC also indicates the priorities in mitigation actions such as in the energy, oil and gas extraction, and transport sectors. The energy sector is the largest emitter of GHGs for which the country aims to increase the introduction of energy efficiency measures, as well as alternative and renewable energies.

In 2013 and 2014, multilateral and bilateral providers committed about USD 63 million per year to climate actions in the country. This is considerably lower than the average among EECCA countries (i.e. USD 303 million per year) and a similar level to that for Kyrgyzstan. This may reflect the country's high level of economic development (USD 16 710 per capita gross domestic product purchasing power parity [GDP PPP] in 2014 was the third highest after Kazakhstan and Belarus) and less need for development finance.

Multilateral institutions were the dominant channel to deliver climate-related development finance to Azerbaijan in 2013 and 2014, accounting for USD 51 million per year (or 80% of all channels). The major contributors included the WBG, the ADB and the EBRD. The largest amount of climate-related development finance was committed to the waste management and disposal sector and the transport sector in 2013 and 2014. This is attributed to two large-size projects in these sectors by the World Bank and the Asian Development Bank. Other projects on energy efficiency and renewable energy were committed between 2011 and 2015. Apart from development finance, about 80% of foreign direct investment flows were intended for the oil and gas sector in 2014.

A range of ministries and governmental agencies, as well as domestic public financing mechanisms, engage in climate-related projects that are supported by international sources (Government of Azerbaijan, 2016). For instance, the State Agency for Alternative and Renewable Energy Sources acts as a principal regulatory institution for renewable energy resources. The Ministry of Energy and Industry supervises, regulates and controls the efficient use of the fuel and energy mix, and the State Oil Fund of the Republic of Azerbaijan has invested in a range of infrastructure projects in energy, water supply and sanitation, irrigation systems, and transport, among other sectors.

Figure 4.3. **Climate-related development finance flows committed in 2013-14 (Azerbaijan)**

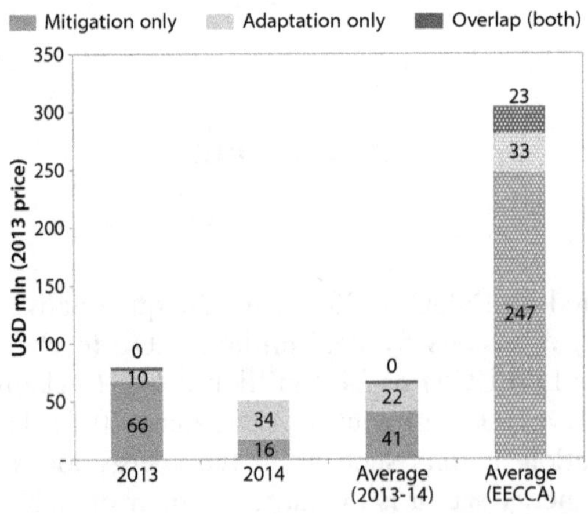

Figure 4.4. **Top 5 sectors in 2013-14**

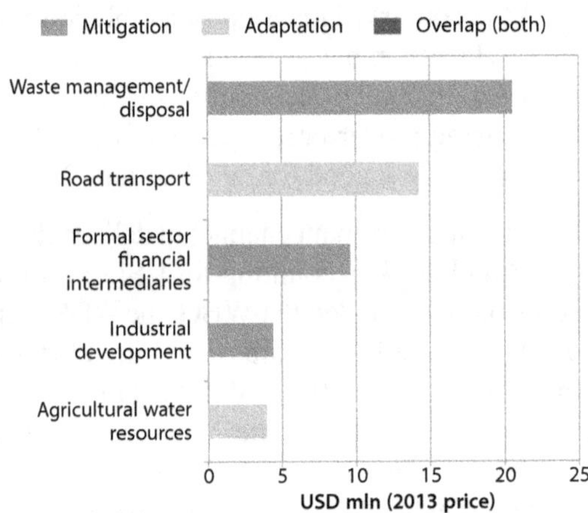

Note: Total climate-related development finance = Mitigation + Adaptation – Overlap (both).

Source: Based on OECD (2016).

Belarus

Belarus submitted its INDC in 2015, communicating its intention to reduce GHG emissions by at least 25% by 2030 below 1990 levels (Government of Belarus, 2015a). While the INDC stresses the importance of adaptation, it notes that specific adaptation measures will be developed in the coming years. Forestry and agriculture are identified as the most vulnerable sectors to climate change. Belarus has already developed a range of legal and policy frameworks for addressing issues concerning climate change and a broader sustainable development agenda (e.g. the State Programme on Mitigation Actions in 2013–20, the National Strategy for Sustainable Development until 2030 and the Concept of the Law on Climate Protection) (Government of Belarus, 2015b).

During 2013-14, USD 140 million per year of climate-related development finance was committed to Belarus, 99% of which was to be provided to mitigation projects. The level of commitment to Belarus is lower than average among the countries of EECCA (USD 303 million per year) during the two-year period. Given that Belarus's GDP per capita PPP is the second highest among the EECCA countries, climate actions including adaptation measures seem to be largely financed by domestic sources. For instance, the average expenditure from Belarus's national and regional budget for energy-saving measures was USD 319 million per year in 2013 and 2014; this was 2.3 times larger than the (international) climate-related development finance committed to the country during the same period.

Multilateral development banks (MDBs) such as the WBG and the EBRD committed the largest amounts of climate-related development finance in 2013-14 (about 96% of total), mainly through non-concessional loans. In the two-year period, the largest amount of financing was directed to the energy sector for projects such as renewable energy development and energy saving. The committed amount to the banking sector was also large. It took the form of a credit line supported by the EBRD for energy efficiency measures called the Belarus Sustainable Energy Finance Facility.

The Ministry of Natural Resources and Environment develops and implements national policies in climate change, both in mitigation and adaptation, and co-ordinates relevant government bodies. This ministry, as well as others such as the Department for Energy Efficiency, Ministry of Energy and Ministry of Forestry, is involved in a range of climate-related projects supported by international sources.

Figure 4.5. **Climate-related development finance flows, committed in 2013-14 (Belarus)**

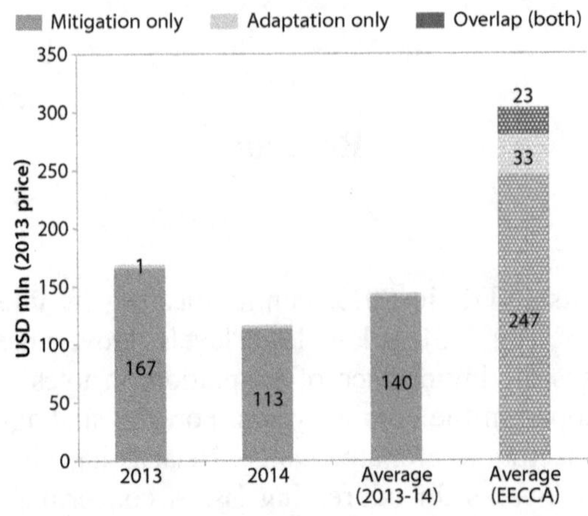

Figure 4.6. **Top 5 sectors in 2013-14**

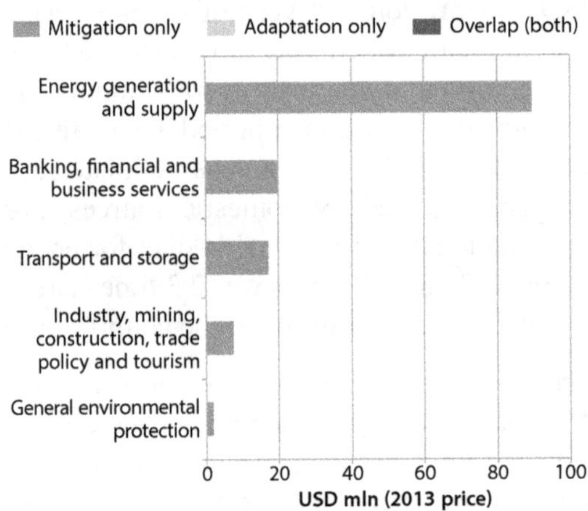

Note: Total climate-related development finance = Mitigation + Adaptation – Overlap (both).

Source: Based on OECD (2016).

Georgia

Georgia submitted its INDC in 2015, highlighting the need for addressing both adaptation and mitigation (Government of Georgia, 2015). Through its INDC, Georgia communicated its intention to reduce GHG emissions by at least 15% below the business-as-usual scenario (BAU) by 2030. Georgia has been developing a range of legal and policy frameworks, relating to climate change and the wider sustainable development agenda (e.g. the Law on Electricity and Natural Gas, and the Low-Emission Development Strategy to be finalised soon) (Government of Georgia, 2016).

During 2013-14, approximately USD 239 million per year of climate-related development finance was committed to support mitigation and adaptation actions in Georgia, but the amounts fluctuated considerably between these two years. The level of the committed amount was lower than average for the EECCA countries (i.e. USD 303 million per year). However, annual climate-related development finance "per capita" committed to the country (approximately USD 55 per capita per year) was about double the EECCA average (USD 27 per capita annually).

The largest amount of climate-related development finance in 2013 and 2014 was committed to the energy sector (i.e. 67.8% of total). Examples of large-scale energy projects include the development or rehabilitation of hydropower plants and energy efficiency in transmission networks. There have been projects on other types of renewable energy (e.g. biomass and wind energies), and energy efficiency on the demand side over the past five years. With regard to adaptation, climate-related development finance was directed mostly to forestry, agriculture and disaster risk management. Most of the climate-related finance committed in 2013 and 2014 was delivered through MDBs (67% from the EBRD and the International Finance Corporation), followed by bilateral donors (32% from the EU, France and Germany). Loans are predominantly used as financial instruments.

Within the country, the Ministry of Environment and Natural Resource Protection is the national focal point, or designated authority, to the UNFCCC, the Green Climate Fund and the Global Environment Facility. It is involved in a range of climate-related projects supported by international sources. However, many other ministries and governmental agencies (e.g. the Ministry of Economy and Sustainable Development), as well as domestic public financing mechanisms also engage in and/or co-finance such projects (Government of Georgia, 2016).

Figure 4.7. **Climate-related development finance flows, committed in 2013-14 (Georgia)**

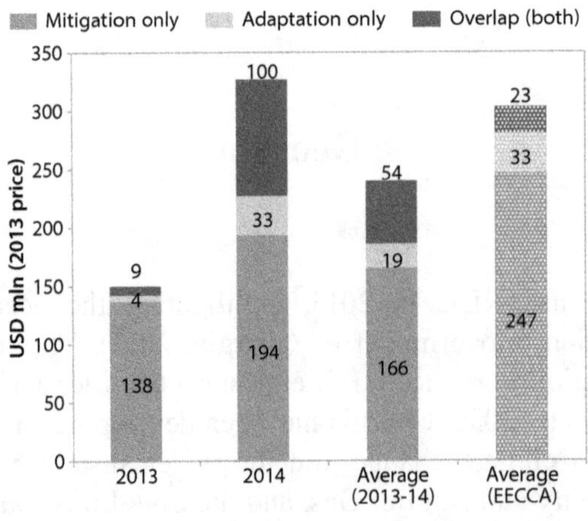

Figure 4.8. **Top 5 sectors in 2013-14**

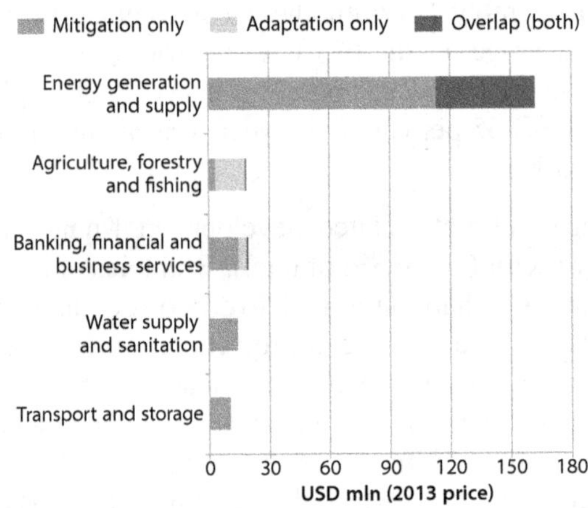

Note: Total climate-related development finance = Mitigation + Adaptation – Overlap (both).

Source: Based on OECD (2016).

Kazakhstan

Kazakhstan submitted its INDC in 2015 with the quantitative target to reduce GHG emissions by 15-25% by 2030 compared to 1990 levels (Government of Kazakhstan, 2015). Kazakhstan has also adopted a range of legal and policy frameworks on addressing issues concerning climate change and a wider set of sustainable development agenda (e.g. Concept for Transition to a Green Economy, the Law on Energy Saving and Energy Efficiency and the Law on Supporting the Use of Renewable Energy Sources).

During 2013-14, about USD 346.7 million of climate-related development finance was committed to Kazakhstan; 91% of the finance targeted mitigation projects. The volume of climate-related development finance committed to Kazakhstan was slightly larger than the average amount in all EECCA countries (USD 303 million/year), while the country's GDP per capita PPP is the highest in the EECCA region. Kazakhstan is also a provider of Official Development Assistance, which is largely directed to Kyrgyzstan, Tajikistan and Ukraine in the region (not limited to climate-related support).

A significantly large share of climate-related development finance was delivered through multilateral channels (USD 311 million per year, or 89.6% of total) in 2013-14. Examples include the EBRD and the European Investment Bank (EIB), using non-concessional loans, and the Climate Investment Funds (CIF) with concessional loans. The EU has provided a significant amount of grant financing.

The largest amount of climate-related finance was directed to the energy generation and supply sector. It was committed to energy efficiency and renewable energy projects, and aligned with national policies on promoting renewable energy and energy efficiency. The finance for the banking and financial sector mostly represents the extension of credit-lines by the EIB to local banks, aiming to help small and medium-sized enterprises finance mainly renewable energy or energy efficiency measures on the demand side.

The Ministry of Energy is a lead ministry for energy policy and governance, as well as climate policies which were under the Environment Ministry until 2014. The Ministry of Energy is involved in a range of climate-related projects supported by international sources. However, many other ministries and governmental agencies, as well as domestic public financing mechanisms, also co-finance and/or engage in such projects. For instance, the Sovereign Wealth Fund "Samruk-Kazyna" has significantly invested in a number of projects on renewable energy, among others (Government of Kazakhstan, 2016). The Ministries of Agriculture, Investment and Development, and National Economy, among others, have also actively engaged in climate actions.

Figure 4.9. **Climate-related development finance flows, committed in 2013-14 (Kazakhstan)**

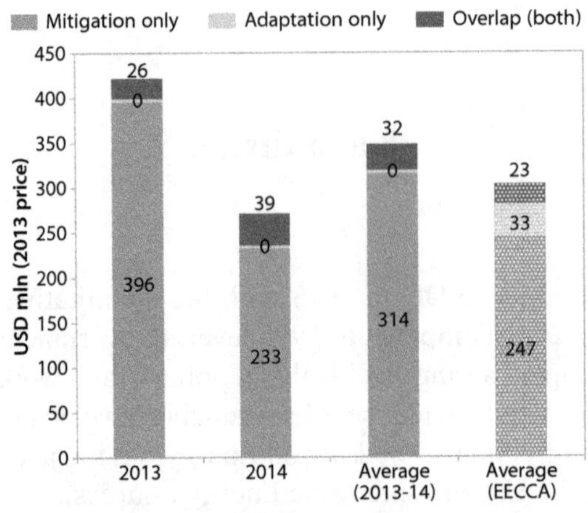

Figure 4.10. **Top 5 sectors in 2013-14**

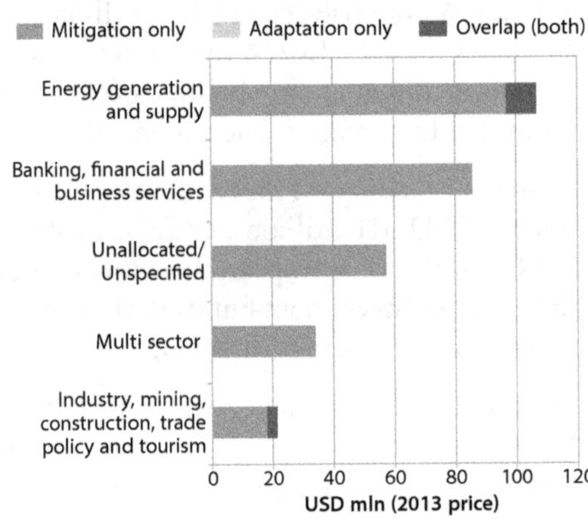

Note: Total climate-related development finance = Mitigation + Adaptation – Overlap (both).

Source: Based on OECD (2016).

Kyrgyzstan

Kyrgyzstan submitted its INDC, outlining both adaptation and mitigation targets and actions. On climate change adaptation, the INDC refers to the "Priorities for Adaptation to Climate Change in the Kyrgyz Republic till 2017". Its priority sectors in adaptation include agriculture, energy, water, emergencies (e.g. disaster risk management), healthcare and forest and biodiversity. Kyrgyzstan has also communicated mitigation targets to reduce GHG emissions by between 11.49% and 13.75% below BAU levels in 2030. Kyrgyzstan has also pledged to reduce GHG emissions by between 29.00% and 30.89% below BAU levels in 2030, contingent on international support (e.g. finance, technology and capacity building) (Government of Kyrgyzstan, 2015).

In 2013-14, bilateral and multilateral donors committed USD 59.9 million per year to climate actions in Kyrgyzstan. This amount of climate-related development finance is five times lower than average amounts for the EECCA countries. The level of finance is lower than that for Tajikistan (USD 260 million/year) and Moldova (USD 136 million/year) whose income levels are similar to that of Kyrgyzstan. The committed finance "per capita" (USD 10.3 per person) is also considerably lower than the EECCA average (USD 33.2 per person).

The largest amount of climate-related development finance was committed to the energy sector (i.e. energy generation and supply) in 2013 and 2014. The banking, financial and business services sector received the second largest amount committed (notably for the Kyrgyzstan Sustainable Energy Financing Facility). Despite the importance of adaptation in the Kyrgyzstan's INDC, the thematic balance in the climate-related development finance between adaptation and mitigation was not well struck in 2013 and 2014: the shares of committed amounts to mitigation, adaptation and multi-focal projects were 69%, 10% and 21%, respectively.

The Coordinating Commission on Climate Change is responsible for ensuring multi-sector co-ordination of all climate actions in Kyrgyzstan. It consists of all relevant ministries and divisions, and representatives of the civil, academic and business sectors. The State Agency for Environmental Protection and Forestry and the Ministry of Energy and Industry also play key roles in developing and implementing climate- and energy-related policies. International sources financed 97% of public investments (including in climate-related ones) in 2012 (PPCR, 2015).

Figure 4.11. **Climate-related development finance flows, committed in 2013-14 (Kyrgizstan)**

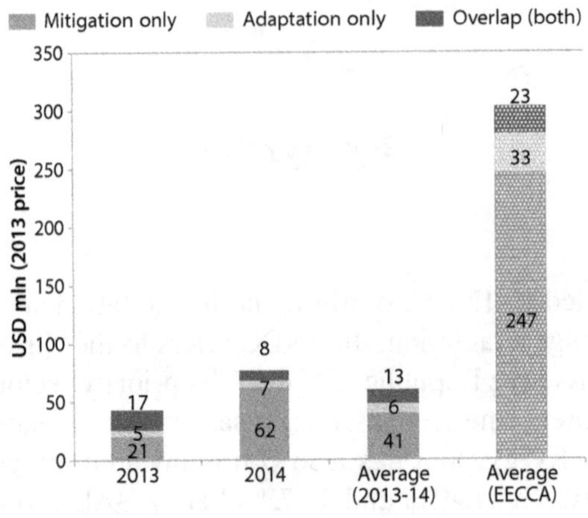

Figure 4.12. **Top 5 sectors in 2013-14**

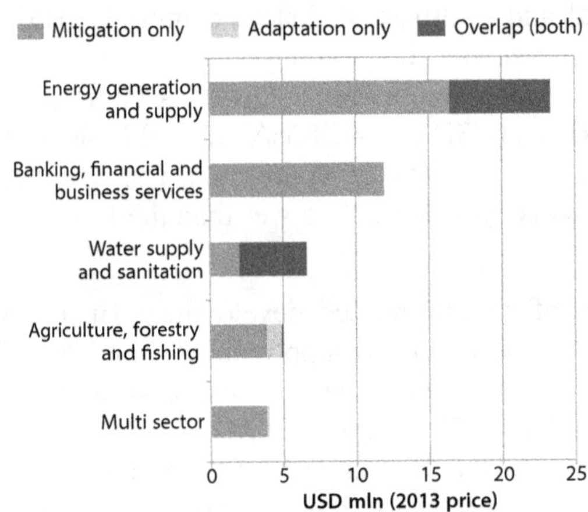

Note: Total climate-related development finance = Mitigation + Adaptation – Overlap (both).

Source: Based on OECD (2016).

Moldova

The Republic of Moldova (Moldova) submitted its INDC in 2015, highlighting the need for addressing both adaptation and mitigation. Through its INDC, the country communicated its intention to reduce its GHG emissions by 64-67% below its 1990 level by 2030, and to reduce GHG emissions by 78% by 2030 if international support is available (Government of Moldova, 2015). Moldova's Low-Emission Development Strategy (LEDS) has been developed and plans were to be approved in 2016. The LEDS outlines key steps on climate actions for the period up to 2030. The INDC also refers to Moldova's Climate Change Adaptation Strategy until 2020 and the Action Plan on its implementation, in order to outline its mid-term adaptation vision, goal and targets.

During 2013-14, USD 136.2 million of climate-related development finance was committed to Moldova – 68% for mitigation, 19% for adaptation and 11% to multi-focal projects (both mitigation and adaptation). The finance is substantially smaller than the average commitments to the 11 EECCA countries (USD 303 million/year). However, the committed finance "per capita" (USD 38 per person) is slightly higher than the average in EECCA countries (USD 33 per person).

The largest amounts of climate-related development finance in 2013 and 2014 were committed to the energy sector (generation and supply) and the agriculture, forestry and fishery sectors. The latter was committed almost the same amounts in adaptation and mitigation activities. While adaptation and multi-focal projects were committed at a level similar to that of the EECCA average, the finance committed to mitigation projects was substantially lower than average. About 80% of climate-related development finance flow was committed through multilateral channels, using mainly loans (EBRD, EIB and WBG); bilateral sources (EU, Germany and Japan) committed the reminder, mainly in the form of grants.

The National Commission for the Implementation and Realization of the Commitments implements and achieves the commitments under the UNFCCC (chaired by the Minister of Environment). A number of other ministries (e.g. Ministry of Environment, Ministry of Economy and Ministry of Regional Development and Construction, Ministry of Agriculture and Food Industry) and domestic financial mechanisms (the Energy Efficiency Fund and the National Fund for Regional Development) have also engaged in climate-related projects, including public investment (Government of Moldova, 2016).

Figure 4.13. **Climate-related development finance flows, committed in 2013-14 (Moldova)**

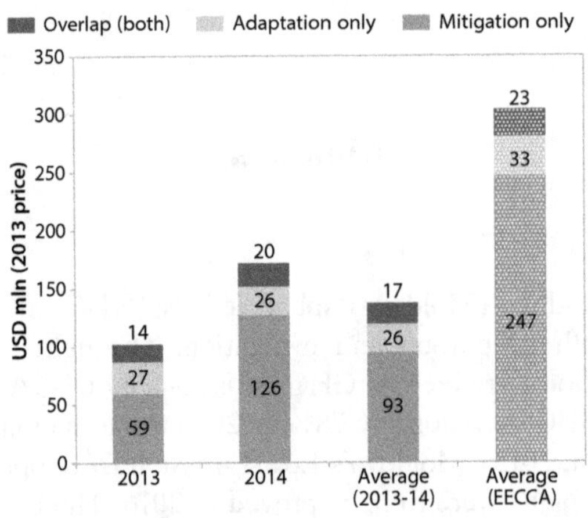

Figure 4.14. **Top 5 sectors in 2013-14**

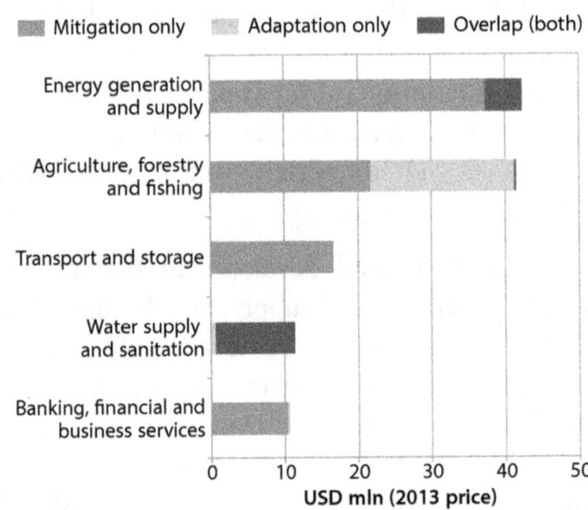

Note: Total climate-related development finance = Mitigation + Adaptation – Overlap (both).

Source: Based on OECD (2016).

Tajikistan

The Republic of Tajikistan (Tajikistan) submitted its INDC in 2015, communicating unconditional and conditional (on international support) targets with regard to both adaptation and mitigation. The unconditional mitigation target is not to exceed 80-90% of GHG emissions at the 1990 level by 2030, whereas the conditional target is not to exceed 65-75% on the same basis (Government of Tajikistan, 2015).

In 2013-14, USD 286 million per year of the climate-related development finance was committed to Tajikistan. This amount is slightly smaller than the EECCA average (i.e. USD 303 million per year), while the committed finance "per capita" (USD 31.3 per person) is also slightly lower than average (USD 33.2 per person). Most of the finance was committed in the form of either grants or concessional loans, reflecting the relatively low level of economic development.

While 61% of finance was committed to mitigation projects, 20% was committed to multi-focal projects (both mitigation and adaptation), most of which was for two large-scale activities in the energy and agriculture sectors. MDBs, bilateral donors and climate funds all committed significant amounts of climate-related development finance in 2013 and 2014. Major contributors of climate-related development finance during the period included the ADB, EBRD, Climate Investment Funds (CIF), WBG, Germany and Switzerland.

Tajikistan is the first country in the EECCA region to participate in the multi-donor Pilot Program for Climate Resilience (PPCR) managed by the CIF. The PPCR includes sub-projects such as enhancing resilience of the energy sector, improving rural livelihood and land use, and supporting small and medium-sized enterprises/farmers. Nonetheless, the energy sector was committed by far the largest amount of climate-related development finance during 2013-14 (i.e. about USD 170 million per year, or 67% of the total).

While the Committee on Environmental Protection is responsible for natural resources management and environmental protection, other ministries and governmental bodies are also involved in climate-related projects and programmes. In general, the share of financial support from international sources in public investments remains considerably high in Tajikistan; of the USD 2.13 billion of public investment within the country from 2002-12, only USD 147 million came from the government budget.

Figure 4.15. **Climate-related development finance flows, committed in 2013-14 (Tajikistan)**

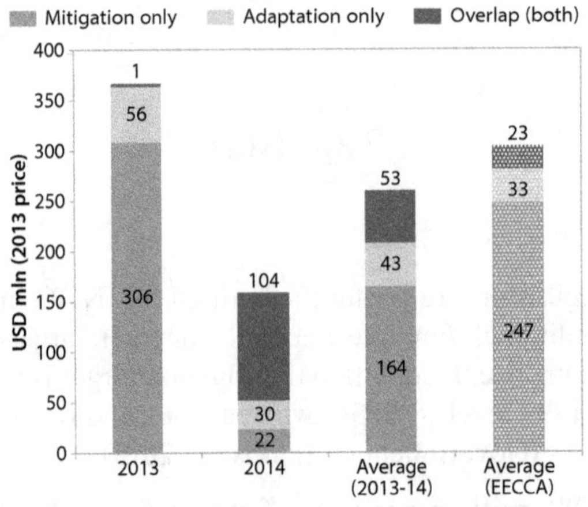

Figure 4.16. **Top 5 sectors in 2013-14**

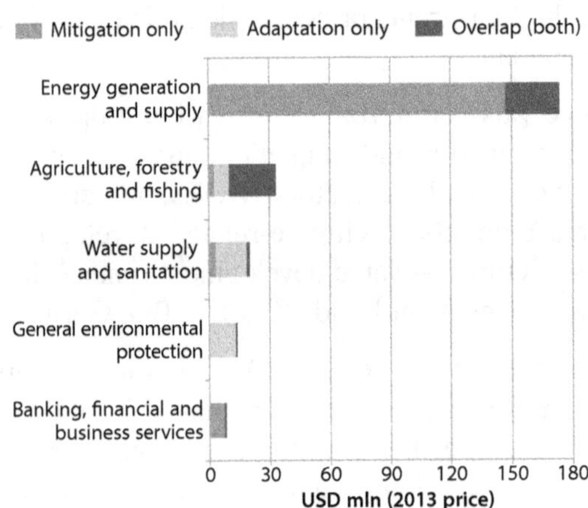

Note: Total climate-related development finance = Mitigation + Adaptation – Overlap (both).
Source: Based on OECD (2016).

Turkmenistan

Turkmenistan's INDC, submitted in 2015, has communicated both conditional (on international support) and unconditional mitigation targets. Through the unconditional target, the growth rate of GHG emissions will be lower than the growth rate of the country's GDP between 2015 and 2030. During that same period, the conditional target is that GHG emissions will not increase. The INDC also stresses the importance of preparing a detailed national action plan for adaptation. Agriculture, water management, health, soil and land resources, ecosystems (flora and fauna) and forestry are identified as the most vulnerable sectors to climate change (Government of Turkmenistan, 2015).

Turkmenistan receives a considerably small size of climate-related development finance, compared with amounts committed to other EECCA countries. In 2013-14, the committed financial flows to Turkmenistan from international sources amounted to USD 3.3 million per year to mitigation projects, and USD 1.6 million per year to adaptation projects – merely 1.6% of the EECCA average (i.e. USD 303 million per year). This reflects the country's view that domestic finance will be the primary source for its climate actions as described in its INDC.

The GEF committed 98% of climate-related development finance to the country in 2013 and 2014. It supported two large-scale projects, one on energy efficiency and renewable energy for water management systems (which could have also fallen into the water sector) and one on climate-resilient livelihoods in agricultural communities.

The major policies of Turkmenistan to mitigate climate change are reflected in the main government programmes, especially in the National Strategy of Social and Economic Transformation of Turkmenistan until 2030 and the National Strategy of Turkmenistan on Climate Change. The latter also includes possible measures for energy efficiency. The government is developing NAMAs. By the end of 2014, no binding target on energy efficiency or renewable energy had been established, although the INDC includes targets on GHG emissions (i.e. not to increase GHG intensity towards 2030). Adaptation to climate change is also a major focus of the National Strategy on Climate Change. The strategy will be implemented through the National Action Plan for Adaptation, which is under development and meant to become an integral part of the national programmes and plans for socio-economic development.

Figure 4.17. **Climate-related development finance flows, committed in 2013-14 (Turkmenistan)**

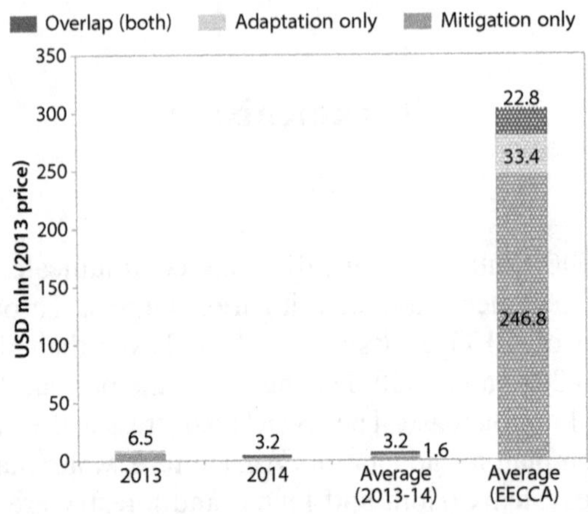

Figure 4.18. **Top 5 sectors in 2013-14**

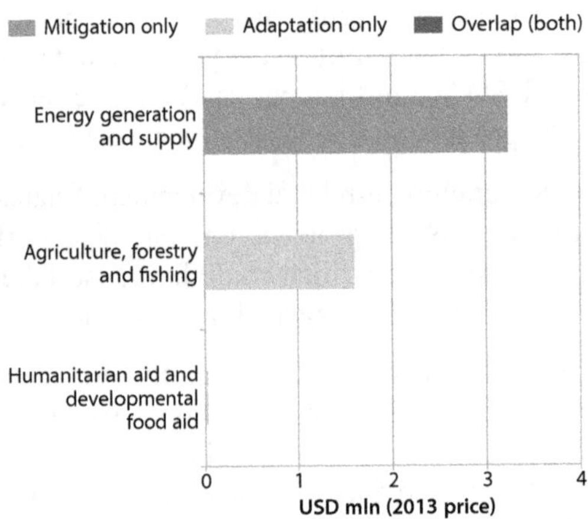

Note: Total climate-related development finance = Mitigation + Adaptation – Overlap (both).

Source: Based on OECD (2016).

Ukraine

Ukraine submitted its INDC with a target of reducing GHG emissions by at least 40% below 1990 levels by 2030, including LULUCF. The INDC does not indicate any adaptation-related targets or actions, but states it will consider adaptation with the same priority as mitigation "for a medium-term outlook" (Government of Ukraine, 2015).

Nearly USD 860 million per year of climate-related development finance has been committed to Ukraine in 2013 and 2014. This amount is significantly larger than average among the EECCA countries (i.e. USD 303 million per year), and the second largest only after Uzbekistan. Nevertheless, annual climate-related development finance committed to Ukraine (approximately USD 19 per capita/year) is about 30% smaller than the EECCA average (USD 27 per capita/year).

Nearly 90% of climate-related development finance was committed to the energy generation and supply sector in 2013 and 2014. A number of donors and financial institutions committed a significant amount of finance to energy efficiency and renewable energy related projects, both on the supply and demand sides. This reflects the need to improve the GHG efficiency of Ukraine's energy sector, which is one of the most energy-intensive in the world. The "unallocated and unspecified" sector is the second largest sector supported in the observed period. In this sector, the two largest projects supported by MDBs were the subway system development project and the railway tunnel development, which could have also been recorded as transport sector projects.

Both bilateral and multilateral providers committed significant amounts of climate-related development finance in 2013 and 2014. The EBRD, WBG, EIB and CIF committed the largest amounts of climate-related development finance during the two-year period, mainly using non-concessional loans. Bilateral donors such as Denmark, the EU, Germany, Sweden, Switzerland and the United States also committed large amounts of grants to the country during the period.

The Ministry of Ecology and Natural Resources is in charge of the development and implementation of state environmental policies, including climate change issues. However, many other ministries and governmental agencies, as well as domestic public financing mechanisms, also co-finance and/or engage in such projects. This list includes the State Agency on Energy Efficiency and Energy Saving.

Figure 4.19. **Climate-related development finance flows, committed in 2013-14 (Ukraine)**

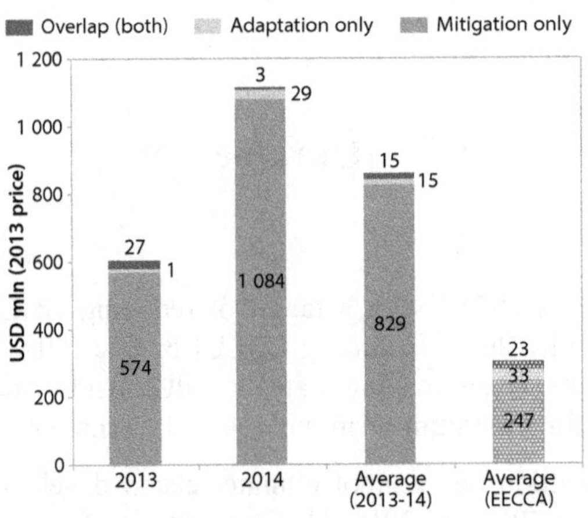

Figure 4.20. **Top 5 sectors in 2013-14**

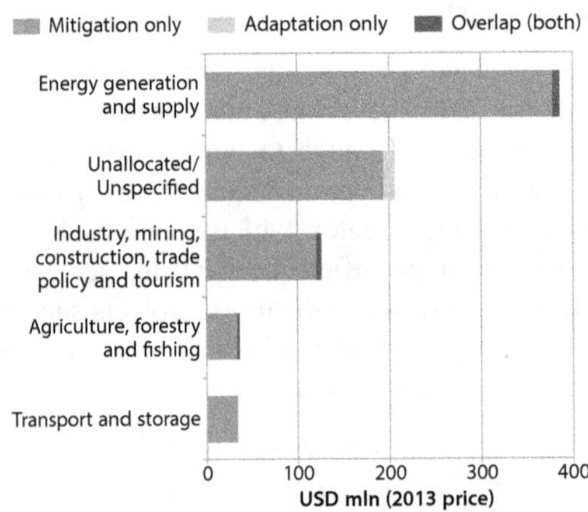

Note: Total climate-related development finance = Mitigation + Adaptation – Overlap (both).
Source: Based on OECD (2016).

Uzbekistan

As of September 2016, Uzbekistan had not communicated its INDC. However, the government has announced it will reduce GHG emissions by promoting renewable energy development and energy efficiency measures, including rational energy pricing (Government of Uzbekistan, 2009). To date, there has been no quantified target for GHG emissions, energy consumption/intensity or installed capacity of renewable energy plants. There is no comprehensive, national-level policy that promotes adaptation, either.

More than USD 1 billion per year of climate-related development finance was committed to mitigation and adaptation projects in Uzbekistan in 2013-14, which was the largest amount among the 11 EECCA countries (the average committed amount to the EECCA countries was USD 303 million per year per country). Japan committed to two large projects on gas-fired power plants, which accounted for 51% of the total committed finance during the period. Nonetheless, even without these two projects, a significantly larger amount of climate-related development finance (about USD 500 million per year) was committed to climate actions in the country. For instance, the agriculture and water sectors together received USD 245 million per year of climate finance, including for a range of adaptation projects.

Both bilateral and multilateral providers committed significant amounts of climate-related development finance in 2013 and 2014. As mentioned above, Japan was the largest contributor during the period, providing USD 540 million per year of concessional loans. Among multilateral channels, the ADB, WBG and Islamic Development Bank committed significant amounts of finance in the form of both concessional and non-concessional loans.

Uzbekistan has mobilised a considerable amount of domestic finance for climate-related projects and for a wider set of development activities. It created the Fund for Reconstruction and Development in mid-2006. Between its creation and 2014, the Fund had accumulated USD 15 billion in assets, most of which were managed by the Central Bank of Uzbekistan. The Fund has also financed several projects supported by international climate-related development finance. Moreover, a centralised electricity system operator "UzbekEnergo" developed its periodic investment plans in energy efficiency implementation, as well as in energy sector infrastructure in general, amounting to USD 5 billion over 2011-15. The State Committee for Nature Protection is responsible for the protection of environment and natural resources, and works with other ministries on climate change-related issues.

Figure 4.21. **Climate-related development finance flows committed in 2013-14 (Uzbekistan)**

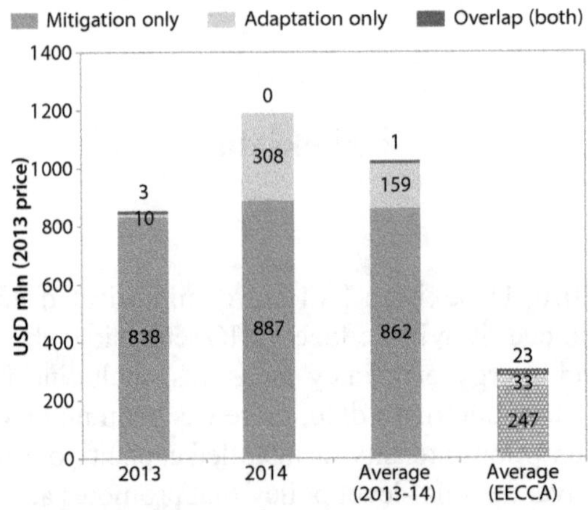

Figure 4.22. **Top 5 sectors in 2013-14**

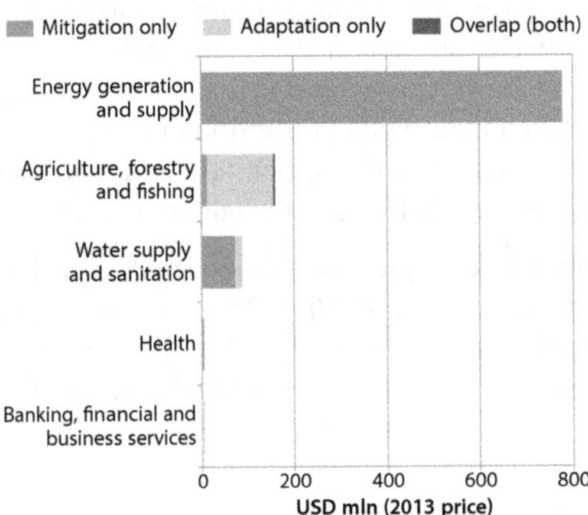

Note: Total climate-related development finance = Mitigation + Adaptation – Overlap (both).

Source: Based on OECD (2016).

Notes

1. Please see: www.oecd.org/env/outreach/eap-tf.htm.
2. For more details, see www.oecd.org/dac/stats/climate-change.htm and on the DAC members see www.oecd.org/dac/dacmembers.htm.

References

Government of Armenia (2016), First Biennial Update Report, Government of Armenia, http://unfccc.int/resource/docs/natc/armbur1.pdf (accessed 24 October 2016).

Government of Armenia (2015), Intended Nationally Determined Contribution of the Republic of Armenia under the UN Climate Change Framework Convention, http://www4.unfccc.int/submissions/INDC/Published%20Documents/Armenia/1/INDC-Armenia.pdf (accessed 24 October 2016).

Government of Azerbaijan (2016), The First Biennial Updated Report of the Republic of Azerbaijan to UNFCCC, Government of Azerbaijan, http://unfccc.int/resource/docs/natc/aze_bur1_eng.pdf (accessed 24 October 2016).

Government of Azerbaijan (2015), To the United Nations Framework Convention on Climate Change (UNFCCC) on the Intended Nationally Determined Contribution (INDC) of the Republic of Azerbaijan, http://www4.unfccc.int/submissions/INDC/Published%20Documents/Azerbaijan/1/INDC%20Azerbaijan.pdf (accessed 24 October 2016).

Government of Belarus (2015a), Intended Nationally Determined Contributions of the Republic of Belarus, http://www4.unfccc.int/submissions/INDC/Published%20Documents/Belarus/1/Belarus_INDC_Eng_25.09.2015.pdf (accessed 24 October 2016).

Government of Belarus (2015b), The Sixth National Communication to the UNFCCC, Government of Belarus, http://unfccc.int/resource/docs/2015/idr/blr06.pdf (accessed 24 October 2016).

Government of Georgia (2015), *Intended Nationally Determined Contribution*, Government of Georgia, Tbilisi http://www4.unfccc.int/submissions/INDC/Published%20Documents/Georgia/1/INDC_of_Georgia.pdf (accessed 24 October 2016).

Government of Georgia (2016), The First Biennial Updated Report of the Republic on Climate Change, Government of Georgia http://unfccc.int/files/national_reports/non-annex_i_parties/ica/application/pdf/first_bur_-_georgia.pdf.

Government of Kazakhstan (2016), Kazakhstan's Biennial Report to UNFCCC, the Government of Kazakhstan, http://unfccc.int/files/national_reports/biennial_reports_and_iar/submitted_biennial_reports/application/pdf/br-text_eng_kz.pdf (accessed 24 October 2016).

Government of Kazakhstan (2015), Intended Nationally Determined Contribution - Submission of the Republic of Kazakhstan, Government of Kazakhstan, http://www4.unfccc.int/submissions/INDC/Published%20Documents/Kazakhstan/1/INDC%20Kz_eng.pdf (accessed 24 October 2016).

Government of Kyrgyzstan (2015), The Kyrgyz Republic Intended Nationally Determined Contribution, Government of the Kyrgyz Republic, http://www4.unfccc.int/submissions/INDC/Published%20Documents/Kyrgyzstan/1/Kyrgyzstan%20INDC%20_ENG_%20final.pdf (accessed 24 October 2016).

Government of Moldova (2016), First Biennial Update Report of the Republic of Moldova under the United Nations Framework Convention on Climate Change, Ministry of Environment of the Republic of Moldova, www.unfccc.int/essential_background/library/items/3599.php?rec=j&priref=7839#beg (accessed 24 October 2016).

Government of Moldova (2015), Republic of Moldova's Intended National Determined Contribution, Government of the Republic of Moldova, http://www4.unfccc.int/submissions/INDC/Published%20Documents/Republic%20of%20Moldova/1/INDC_Republic_of_Moldova_25.09.2015.pdf (accessed 24 October 2016).

Government of the Tajikistan (2015), Intended Nationally Determined Contribution (INDC) towards the achievement of the global goal of the UN Framework Convention on Climate Change (UNFCCC), Government of the Tajikistan, http://www4.unfccc.int/submissions/INDC/Published%20Documents/Tajikistan/1/INDC-TJK%20final%20ENG.pdf (accessed 24 October 2016).

Government of Turkmenistan (2015), Intended nationally determined contribution of Turkmenistan in accordance with decision 1/CP. 20 UNFCCC, the Government of Turkmenistan (GoT), http://www4.unfccc.int/submissions/INDC/Published%20Documents/Turkmenistan/1/INDC_Turkmenistan.pdf (accessed 24 October 2016).

Government of Ukraine (2015), Intended Nationally-Determined Contribution (INDC) of Ukraine to a New Global Climate Agreement, Government of Ukraine, http://www4.unfccc.int/submissions/INDC/Published%20Documents/Ukraine/1/150930_Ukraine_INDC.pdf (accessed 24 October 2016).

Government of Uzbekistan (2009), the Second National Communication of the Republic of Uzbekistan under the United Nations Framework Convention on Climate Change, the Government of Uzbekistan, http://unfccc.int/resource/docs/natc/uzbnc2e.pdf (accessed 24 October 2016).

Nachmany, M. et al. (2015), The 2015 Global Climate Legislation Study: Uzbekistan, Excerpt from the 2015 Global Climate Legislation Study A Review of Climate Change Legislation in 99 Countries, The Grantham Research Institute on Climate Change and the Environment, www.lse.ac.uk/GranthamInstitute/wp-content/uploads/2015/05/UZBEKISTAN.pdf (accessed 24 October 2016).

OECD (2016), "Project-level Data for every Climate-related Development Finance Project in 2013-14", *Climate Change: OECD DAC External Development Finance Statistics* (database), www.oecd.org/dac/stats/climate-change.htm (accessed 24 October 2016).

PPCR (2015), *Aide Memoire: Kyrgyz Republic – Pilot Program for Climate Resilience Joint Multilateral Development Bank Scoping Mission*, Pilot Program for Climate Resilience (PPCR) of the Climate Investment Funds, https://www-cif.climateinvestmentfunds.org/sites/default/files/meeting-documents/2015-10-26_kyrgyz_republic_ppcr_scoping_mission_draft_aide_memoire_final_-_en.pdf (accessed 24 October 2016).

R2E2 (2014), "Scaling Up Renewable Energy Program (SREP): Investment Plan for Armenia", The Energy Saving and Renewable Energy Fund (R2E2), Minister of Energy and Natural Resources of Republic of Armenia, www.r2e2.am/wp-content/uploads/2014/08/Armenia-SREP_2014.pdf (accessed 24 October 2016).

[illegible] (2013), [illegible] *Mainstream* [illegible] *Programme* [illegible] Climate [illegible] of the [illegible] (accessed 24 October 2016).

[illegible] (2011), Scaling Up Renewable Energy Program (SREP) Investment Plan [illegible] for Energy Saving and Renewable Energy Fund [illegible] Minister of [illegible] and Natural Resources of Republic of Armenia, [illegible] (accessed 24 October 2016).

ORGANISATION FOR ECONOMIC CO-OPERATION AND DEVELOPMENT

The OECD is a unique forum where governments work together to address the economic, social and environmental challenges of globalisation. The OECD is also at the forefront of efforts to understand and to help governments respond to new developments and concerns, such as corporate governance, the information economy and the challenges of an ageing population. The Organisation provides a setting where governments can compare policy experiences, seek answers to common problems, identify good practice and work to co-ordinate domestic and international policies.

The OECD member countries are: Australia, Austria, Belgium, Canada, Chile, the Czech Republic, Denmark, Estonia, Finland, France, Germany, Greece, Hungary, Iceland, Ireland, Israel, Italy, Japan, Korea, Latvia, Luxembourg, Mexico, the Netherlands, New Zealand, Norway, Poland, Portugal, the Slovak Republic, Slovenia, Spain, Sweden, Switzerland, Turkey, the United Kingdom and the United States. The European Union takes part in the work of the OECD.

OECD Publishing disseminates widely the results of the Organisation's statistics gathering and research on economic, social and environmental issues, as well as the conventions, guidelines and standards agreed by its members.

OECD PUBLISHING, 2, rue André-Pascal, 75775 PARIS CEDEX 16
(97 2016 24 1 P) ISBN 978-92-64-26632-2 – 2016

www.ingramcontent.com/pod-product-compliance
Lightning Source LLC
LaVergne TN
LVHW081416110826
845149LV00010B/1765

* 9 7 8 9 2 6 4 2 6 6 3 2 2 *